PROGRESS IN URBAN GEOGRAPHY

Also edited by Michael Pacione for Croom Helm

Urban Problems and Planning in the Developed World (1981)
Problems and Planning in Third World Cities (1981)
Progress in Rural Geography (1983)

PROGRESS IN URBAN GEOGRAPHY

Edited by Michael Pacione

CROOM HELM
London & Canberra

BARNES & NOBLE BOOKS
Totowa, New Jersey

© 1983 Michael Pacione
Croom Helm Ltd, Provident House, Burrell Row,
Beckenham, Kent BR3 1AT

British Library Cataloguing in Publication Data

Pacione, Michael
    Progress in urban geography.
    1. Cities and towns
    I. Title
    910'.091732   GF125

    ISBN 0-7099-2027-X

First published in the USA 1983 by
Barnes & Noble Books
81 Adams Drive
Totowa, New Jersey, 07512

ISBN 0-389-20357-2

Typeset by Leaper & Gard Ltd, Bristol
Printed and bound in Great Britain
by Billing and Sons Ltd Worcester

CONTENTS

FIGURES

TABLES

TO EMMA VICTORIA

PREFACE

Over the last twenty-five years urban geography has emerged as a major focus for research and teaching in human geography. This is due partly to the fact that a substantial and increasing proportion of the world's population now lives in towns and cities; partly to the recent and growing interest in the problem-solving or applied approach to geography; and partly to the eclectic nature of a sub-discipline with strong linkages to related fields in sociology, economics, psychology, politics and planning. Urban geography is concerned with the operation and effects on the city of a wide range of economic, social and political processes, each of which has generated a field of systematic investigation in its own right. It can be argued that the roots of the subject are to be found more in regional studies than in any systematic tradition.

Contemporary urban geography thus involves both studies of cities as regions and studies of specific themes within the urban setting, such as retailing, transport, housing or health. But urban based investigations are not merely micro-scale applications of some wider perspective — the urban scale of reference brings with it new conceptual and methodological questions and presents unique problems for investigation.

The breadth of the subject and the scale and pace of change make it difficult for scholars to maintain contact with recent developments, keynote statements and relevant articles spread across a wide range of professional journals and less accessible reports. This volume is a direct response to the need for a text which reviews the progress and current state of the subject and which provides a reference point for future developments in urban studies. This collection of original essays is designed to encapsulate the major themes and recent developments in a number of areas of central importance in urban geography.

Michael Pacione
University of Strathclyde
Glasgow

INTRODUCTION

Housing occupies a major part of any urban area. In Chapter 1 David Kirby traces the development of geographical interest in housing research, and identifies demand-based and supply-based explanations. The former, focusing on the competition between households for urban land, incorporates three distinct lines of enquiry — the neo classical economic approach; the essentially descriptive ecological approach; and the behavioural approach which is seen as a direct response to the failure of the more traditional perspectives to explain the spatial patterns identified. All three paradigms, however, fail to give explicit consideration to the constraints placed on demand; particularly important are those imposed by agencies responsible for the supply and allocation of housing. This supply-based perspective is developed in a discussion of the managerialist approach in which particular attention is given to the gatekeeper role of estate agents, private and public landlords, developers and builders, financial institutions, and government. Despite the range of research over the past decade the causal relationships and the processes operating within the housing system are imperfectly understood. Themes which require further empirical investigation include detailed studies of individual areas to determine the relative influence of the various factors within the housing system; the relationship between household choice and the constraints imposed by supply agencies; the identification of sub markets, including minority group needs; spatial variations in house prices; and the effect of public policy on the housing system.

The economic and social problems of large cities and of the inner city in particular have generated considerable debate. In Chapter 2 Paul Bull identifies the absolute and relative shifts in the distribution of employment away from the major cities which have characterised many western countries over the last two decades, and describes the changing pattern of unemployment at the inter- and intra-city scales. The phenomenon of decentralisation resulting from either plant transfer or differential rates of growth or closure between central and peripheral locations is then considered in detail. The relative importance of the mechanisms underlying the process is assessed and a number of possible explanations of these spatial trends in manufacturing activity examined. Given the powerful forces in the economy for manufacturers

1

to substitute capital for labour and to locate away from the inner areas of cities it is difficult to be optimistic about the short to medium term future of inner urban areas. The recent inner city partnerships between central government and local authorities, and the creation of enterprise zones are considered likely to be little more than palliatives. It is concluded that as the chances of creating many new jobs in manufacturing in any Western country are extremely small, the possibility of a major reduction in the current high levels of unemployment in the major cities appears to be slight.

Investigation of social problems is a prime concern of urban geographers and David Herbert's discussion of crime and delinquency in Chapter 3 is firmly embedded in this applied tradition. He first traces the emergence of a 'geography of crime' from its roots in cartographic criminology before considering the data collection and definitional difficulties facing researchers, including changes in the definition of offences and in the spatial basis of statistics, and the low correspondence of police areas with either census or administrative areas. A number of criminological theories are then examined and it is suggested that the flexibility of the subcultural concept which suggests the existence of identifiable delinquent groups commends its adoption in preference to many other positivist stances. Discussion of recent theoretical and methodological developments is followed by an assessment of several approaches to crime and delinquency adopted by geographers. Particular attention is given to the geographical study of crime and delinquency areas in cities; to the question of the emergence and persistence of problem areas; and to the renewed interest in offence patterns and the concept of vulnerable areas. It is suggested that while a geography of crime which focuses on neighbourhood effects is useful recent interest in managerial roles is more likely to lead to explanations for 'crime areas'. The scope for further geographical study of urban crime and delinquency is great with the general aim being reform of the environments which give rise to the problems.

In Chapter 4 Ceri Peach defines the complex concept of ethnicity as the product of two separate structures — the biological and the cultural — with different combinations of the two factors producing a series of distinct ethnicities. In considering the process of assimilation by which ethnicity is modified, he emphasises that the main difference between the spatial and aspatial viewpoints is the importance attached to segregation; the former viewing it as an important causal factor and the latter merely as a casual outcome. He demonstrates how understanding of ethnicity and assimilation has been furthered by studies employing the 'index of dissimilarity' and the construction and use of this device is examined in detail. The relationship between ethnicity and class and the effect of income on segregation are then discussed before attention turns to the critical question of the degree to which ethnic segregation is the result of choice or constraints imposed by a hostile environment. It is concluded that while the positivist paradigm has greatly advanced knowledge it fails to recognise that ethnicity is a transactional or layered rather than categorical phenomenon. It is suggested that it may be from within this less clearly defined area of enquiry that the next impetus for the study of urban ethnicity will emerge.

In considering the key question of urban government and finance in Chapter 5 Ron Johnston examines the evolution, nature and functions of local government in the UK and the USA as examples of systems characterised by markedly different levels of local autonomy. The operations of the State in capitalist society are viewed as being directed towards the twin goals of accumulation (essentially providing the physical and social environment for economic success) and legitimation (ensuring concensus support for the economic system), and urban local governments are an integral part of this system. Attempts to rationalise UK administrative boundaries which have evolved over the preceding century as well as the less centrally directed modifications to the pattern in the USA are examined and the social and financial implications of such changes are discussed. It is concluded that as a result of the State's pursuit of its accumulation goal the 'fiscal squeeze' experienced by local government in many countries may be a permanent feature of late capitalism. Furthermore, since this objective also means promoting efficiency and thus reducing local democracy, urban local government is likely to be the focus of the central-local or efficiency-democracy struggle of the future.

As a consequence of the urban demographic and economic changes of the second half of the 1970s the retail industry presently stands at a crossroads with the expanionist era of the 1960s and 1970s, characterised by large scale activities and radical change in shopping behaviour, being replaced by new limits to growth, the impact of which is as yet only imperfectly understood. In Chapter 6 Ross Davies combines a review of literature dealing with recent changes in urban retailing with an assessment of the likely concerns of geographical enquiry in the future. He foresees a continued diminution of the effective purchasing power of the population as a whole with a continuing dispersal of population to the urban periphery and an increasing threat to the relative health of central city areas. These trends suggest the need for research into the long term future for suburban superstores and purpose-built shopping centres. The socio-economic problems surrounding retailing, such as the difficulties experienced by small independent shops and the lack of access to essential services for people living in isolated communities seem to be of particular concern. He also addresses the question of equity versus efficiency in provision of retail facilities from the viewpoint of consumer, trader and planner; and considers the likely changes in the structure of retail employment, before speculating on the potential impact of technological change on the future pattern of urban retailing.

Transport is the 'life-blood' of cities. In Chapter 7 Peter White analyses the question of urban travel from both the demand and supply viewpoints. He first considers the role of factors such as trip length, travel mode, and the effects of city size on the structure of urban transport demand. Differences in travel behaviour related to personal and household characteristics, and the change in movement patterns over time are also discussed before the roles of public and private transport in Britain are compared with the situation in other countries. On the supply side of the equation attention is given to finance and pricing policies for urban transport and the concept of system management is explored in detail. The effects of technological developments on future urban transport are assessed and likely changes in trip patterns suggested. It is concluded that the theoretically attractive objective of lessening pressure on the transport system by reducing the need for movement between home and workplace is unlikely to be realised in the foreseeable future. The more realistic prospect is for a similar volume of traffic spread more evenly through the day, with the public transport share being heavily dependent upon the cost and quality of service offered.

The urban environment has often been indicted as having an adverse effect upon human health and behaviour. In Chapter 8 John Giggs reviews the main conceptual approaches and empirical research undertaken by medical geographers at the intra-urban scale. Four major areas of enquiry are examined: disease mapping, ecological associative analysis, disease diffusion studies, and the geography of health care, with particular attention devoted to the latter two themes. Despite practical problems, mapping the incidence of ill health and mortality can have a significant explanatory value, especially if the revealed patterns are related to environmental conditions. A major difficulty for associative analyses, however, is that the effects of the environment on health are normally subject to considerable time lag. This factor seriously complicates the interpretation of static or 'period picture' investigations.

It is suggested that the range of techniques used in diffusion studies may offer a superior insight into many medical phenomena. Considerable effort has recently been invested in the study of the spatial aspects of health service delivery systems and their utilisation patterns. Three major research areas within this field are subjected to detailed examination. These refer to the supply components of the medical care system (for example, doctors and hospitals) and their locational determinants; behavioural studies of the 'consumers' of health care services; and planning oriented research in which the chief interest is in resolving the mismatch between supply and demand for specific health services.

The question of equity in the allocation of services and public facilities in cities has emerged as a major focus of concern in urban geography. In Chapter 9 Andrew Kirby and Steven Pinch discuss the origins of geographical studies of service provision and examine the major conclusions to emerge from empirical research, before critically examining explanations for variations in levels of service provision based on the major theoretical perspectives of pluralism, managerialism and structuralism. They underline the need for closer interaction between those concerned with the essentially spatial aspects of public facility location and those investigating welfare oriented services via broader social theories which attempt to explain patterns of resource allocation. Attention is then given to the definition of public services and the concept of need. In the latter case the definitional problem is compounded by the ecological fallacy and lack of information on the effects of services on individual well-being. Several themes for future research are identified including detailed examination of particular services in order to derive better indices of needs, service quality and outcomes; and expanding the range of phenomena studied to achieve better understanding of the distribution of real income in the city. It is concluded that a major obstacle to attaining the prime objective of an adequate explanation for spatial variations in service provision and well being is the multitude of factors and agencies at work, and that a useful starting point is to recognise the underlying importance of scale.

Pollution is an inevitable problem for all societies. It is the result of an imbalance in the importance attached to economic growth on the one hand and environmental quality on the other, and is most evident in urban areas where man's waste-forming activities are concentrated. In Chapter 10 Derek Elsom discusses the nature of pollution and pollutants in general and then gives particular attention to the major urban problem of air pollution. The growth, measurement, incidence and underlying causes of this problem are examined with reference to the situation in several countries. Switching from problem definition to possible solutions the development of approaches to pollution control in the UK, EEC and USA and in several socialist states is subjected to detailed analysis. Such international comparisons reveal the different emphases placed on pollution control and demonstrate the fundamental importance of economic and political factors in determining environmental quality. While a positive response to urban pollution is related to a country's general economic climate it should be remembered that some deterioration of the environment may be permanent or at best can involve future generations in costly remedial programmes. It is concluded that pollution control programmes in many areas are, unfortunately, still piecemeal and curative rather than comprehensive and preventative.

1 HOUSING

D.A. Kirby

Introduction

According to Robson (1979, p. 67) '"the housing problem" has only come within the purview of geography very recently'. While the validity of this view is open to debate, it is certainly true that until the early 1970s, housing was a much neglected aspect of urban geography. Apart from a small number of statistical analyses of particular issues (Hartman and Hook, 1956), early studies tended to be either descriptive examinations of the different types of housing (Fuson, 1964) or studies in which housing featured as an important but secondary issue, the main concern being either the study of urban structure (Jones, 1961; Robson, 1966) or the study of transport patterns (Kain, 1962; Getis, 1969). The former approach was followed mainly by the cultural geographers of the period for whom, as Wagner (1969) has observed, housing held a double fascination by contributing (as the studies by Smailes, 1955 and Conzen, 1960, emphasised, for example), to the distinctive character of landscapes and standing as 'the concrete expressions of a complex interaction among cultural skills and norms, climatic conditions, and the potentialities of natural materials'. Invariably such approaches failed to generate universal laws and theories (Garrison, 1962) and with the advent of the 'new' geography of the 1960s, gradually faded from the forefront of geographical enquiry, although some interesting examples were produced at the end of the decade (Whitehand, 1967; Rapoport, 1969; Grimshaw *et al.*, 1970). In contrast, the latter approach, based on the seminal works of Burgess (1924) and Hoyt (1939), was more characteristic of the nomothetic viewpoint and reflected the geographer's increasing concern with space, spatial relationships and spatial patterns. Despite such studies, housing remained a relatively under-researched area of geographical enquiry with no particular focus to the studies which sporadically emerged, though a series of articles by British urban geographers at the turn of the decade was to herald a shift in geographical interest in, and awareness of, the housing question (Blowers, 1970; Spencer, 1970; Drakakis-Smith, 1971; Kirby, 1971).

Even so, by the mid 1970s, geography could be fairly criticised for 'never having concerned itself deeply with the question of housing' (Kirby, 1976, p. 2). From the late 1960s in America and early 1970s in Britain and Western Europe, however, there had been increasing concern over the preoccupation of society with economic issues and efficiency and the lack of concern for social welfare and equity. Within geography, it resulted in the now well-documented, radical or relevance revolution (Castells, 1977; Harvey, 1973; Peet, 1977 and Smith, 1977) and a shift in the objects, objectives and methods of geographical enquiry. As Kasperson observed in 1971 (p. 13),

> The shift in the objects of study in geography from supermarkets and highways to poverty and racism has already begun, and we can expect it to continue, for the goals of geography are changing. The new men see the objective of geography as the same as that for medicine — to postpone death and reduce suffering.

As a result of such developments in the subject, geographical interest in housing has increased considerably and by the beginning of the 1980s the vast volume of literature being generated from housing research in geography departments throughout the western world made it possible for a geography of housing to be identified (Bourne, 1981).

While very different from earlier studies of housing and residential geography, there is no uniform approach, as Bassett and Short (1980) have recognised. In part this results from the diversity of social theories upon which the studies have been based and most notably from the distinction between Marxist and non-Marxist doctrines. However, it also results from the very complexity of the subject matter. As Bassett and Short (1980, pp. 1–2) point out

> Housing is a heterogeneous, durable and essential consumer good; an indirect indicator of status and income differences between consumers; a map of social relations within the city; an important facet of residential structure; a source of bargaining and conflict between various power groupings and a source of profit to different institutions and agents involved in the production, consumption and exchange of housing.

If anything, housing is even more than this and because of the complicated interrelationships which exist between housing and its environment (however defined) it is understandable, perhaps, that its study is both complex and varied.

Approaches

As there is no uniform approach to the study of housing, so there is no uniform agreement over the broad approaches which have been followed. For instance Bourne identifies eight distinct but overlapping areas of research (Figure 1.1) which 'vary in scale (macro, micro) and in subject matter (demand, supply, policy) as well as in their philosophy and methodology' (Bourne, 1981, p. 10). In contrast Robson (1969, p. 71) suggests that, in addition to the micro economic approaches of economists such as Evans (1973), Ball and Kirwan (1975) or Whitehead (1975), three major research foci can be recognised. These, he suggests, are social ecology, conflict theory and managerialism, and Marxist concepts. Like Bourne, however, he argues that these separate strands have overlapped and the work

> while having proceeded from different origins, has tended to point in similar directions: by emphasising the importance of constraints rather than choice in access to housing: by illustrating the role of conflict rather than consensus in the goals and interests of the groups involved: and . . . by arguing that class interests lie at the base of much of the system through which housing is produced and distributed. (ibid., p. 71)

Table 1.1: Four Approaches to Housing and Residential Structure

Approach	Wider social theory	Areas of enquiry	Exemplar writers
1. Ecological	human ecology	spatial patterns of residential structure	Burgess (1925)
2. Neo-classical	neo-classical economics	utility maximisation, consumer choice	Alonso (1964)
3. Institutional managerialism locational conflict	Weberian sociology	gatekeepers, housing constraints power groupings, conflict	Pahl (1975) Form (1954)
4. Marxist	historical materialism	housing as a commodity; reproduction of labour force	Harvey (1973, part II) Castells (1977a)

Source: Bassett and Short (1980).

Robson's fourfold classification finds support in the writings of Bassett and Short (Table 1.1) who distinguish between the ecological, the neo-classical, the institutional and the Marxist approaches. While

Figure 1.1: Established Areas of Housing Research

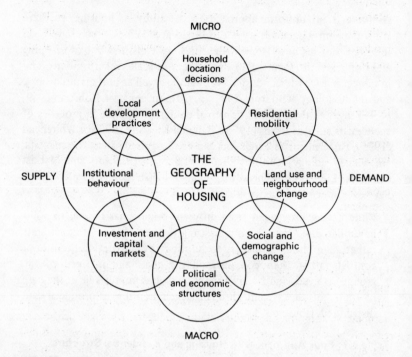

Source: Bourne (1981).

Bourne's eight major research foci could be accommodated within these four major subdivisions, it would seem that the classification suffers from at least one important omission – the behavioural approach to consumer decision-making and residential mobility. This is recognised, in fact, by the authors but they see the behavioural approach as being extra to the scheme, developing out of the ecological and neo-classical approaches. Given the importance of the behavioural approach to the geography of housing (particularly in terms of the volume of research), it is felt that perhaps it warrants individual attention. With the addition of this behavioural category, therefore, it is Bassett and Short's scheme and terminology which will be followed here. As they have observed a 'distinction can be drawn between the earlier developed ecological and neo-classical approaches which focus on equilibrium conditions, housing choices and social harmony and the more recent resurgence of interest in institutional and Marxist approaches which focus on disequilibrium conditions, housing constraints and social conflict' (Robson, 1969, p. 3). Very crudely, this distinction can be equated with, as Bourne suggests, the distinction between those studies of residential differentials focusing upon demand-based explanations and those concerned predominantly with the methods of supply.

Demand-based Explanations

Essentially these studies have focused on the competition between households for land, a location within the city, and a dwelling. They incorporate three distinct lines of investigation — the neo-classical economic approach, the ecological approach and the behavioural approach.

The Neo-classical Economic Approach

Central to this approach is the development of a trade-off model similar to von Thunen's agricultural land-use model. Essentially the basic model has been modified to describe the pattern of land-use within the city. The basis of the approach is the theory that households and firms compete, within the constraints of their budgets, for space within the city, so as to maximise the satisfaction of the various competitors and the efficiency of the urban system. As with the von Thunen model, there are a number of qualifying assumptions which have to be made. First, it is assumed that man acts as a rational economic being and that perfect competition occurs between households and firms. Second, that the urban area is located on an isotropic surface in which transport is equally easy in all directions and transport costs are a direct function of distance. Third, that the Central Business District (CBD) is the most accessible location in the urban system and the only employment centre. Fourth, that the CBD is the most sought-after location, thus producing land prices in the centre which are higher than those in the periphery. In these circumstances, the firms and households compete for space and the residential decision is a 'trade off' between the cost of land and the cost of commuting. As Alonso (1960, p. 154) has observed households, in reaching their decision about where to live, are believed to balance 'the costs and bother of commuting against the advantages of cheaper land with increasing distance from the center of the city and the satisfaction of more space for living'. Under these conditions, the urban rich might be expected to live in large space-consuming properties at low densities in the urban periphery, while the urban poor, unable to pay the high costs of regular commuting, would live in cramped conditions in the inner city.

Although the model seems to fit the general pattern of housing within cities, over the years it has been refined and modified. For instance, the possibility of a multi-centre city has been considered by De Leeuw (1972), while Kain and Quigley (1972 and 1970) have attempted to take account of racial discrimination and variations in the quality of the residential environment. Meanwhile, Pines (1975) has modified the model to allow the rich to live both in the inner city and the outermost suburbs, the assumption being that the former pattern would occur where the higher income groups placed greater premium on time, whereas the latter would occur where space was held to be important. Indeed, with the renovation and rehabilitation of inner city areas, the movement of the urban rich back into the inner city (gentrification) has increased and this is recognised in the modifications proposed by Evans (1973).

As a result of these and various other extensions of, and modifications to, the basic model, valuable insights have been gained into the residential structure of urban areas. However, the models have not helped explain reality and partly because of this, they have tended to decline in popularity. This trend has been stimulated by the criticisms of the model-building approach in general and the trade-off model in particular. It is now generally accepted, for instance, that perfect competition and rational economic man rarely, if ever, exist and, more specifically, it is argued that accessibility costs are of secondary importance and that for most income groups, the availability of a mortgage is paramount (Richardson, 1971). This is related to what is probably the most serious criticism of the neo-classical models, namely their failure to consider the supply and allocation of housing. Under the assumptions of the neo-classical approach, supply is a response to demand and with, perhaps, only one notable exception (Muth, 1969) the major treatise follow this line of reasoning. Even so, despite the weaknesses of the neo-classical economic approach, it has formed an important focus for the study of residential geography and has stimulated research most notably into the development of spatial interaction models. A review of this work is provided in Senior (1974) but for the present it is sufficient to point out that essentially it involves the application of the concepts of Newtonian physics to spatial interaction. Perhaps the basis of the work is a model developed by Lowry (1964) who suggested that the location of employees around their place of work is a function of journey to work costs such that

$$T_{ij} = gE_jf(c_{ij})$$

where T_{ij} = the number of people living in zone i and working in zone j

E_j = the number of jobs in zone j

c_{ij} = the cost of travel from i to j

f = a decreasing function

g = a constant.

This simple gravity-type model has been extended and modified on several occasions. For instance, Wilson (1974) has pointed to the fact that the model implicitly assumes that people and houses are identical and has suggested that the basic Lowry model could be modified to allow for differences in income and house type so that

$$T_{ij}^{kw} = A_i^k B_j^w H_i^k E_j^w f_w(C_{ij})$$

where T_{ij}^{kw} = the number of w income people living in a type k house in zone i, working in zone j

H_i^k = the number of type k houses in each zone i

E_j^w = the number of income w jobs in zone j

$f_w(C_{ij})$ = a function of the cost of travelling from zone i to zone j which varies with income w

A_i^k = a measure of the attractiveness of a type k house in zone i

B_j^w = a measure of the attractiveness of zone j for w income groups.

Indeed, one of the advantages of this approach, as Wilson sees it, is the fact that the spatial interaction model is easily extendable. In his view, such models make up for the fact that with the neo-classical approach 'no very exciting operational models have been developed from the work. Many insights and qualitative analyses have been obtained but few effective models' (ibid., p. 197). What it is important to recognise is that similar criticisms can be levelled at the spatial interaction approach. For instance, the results of an empirical study of nine such models (Openshaw, 1976) have raised 'Severe doubts about the empirical acceptability of current spatial interaction models' (ibid., p. 40) and has pointed to the fact that the theoretically-based models 'are often incapable of operational use because of their excessive data requirements'. However, perhaps the main criticism of this approach is that it is based on a far too simplistic view of the housing market. As Senior (1973, p. 194) observes 'they (the models) need relating to rigorously argued and comprehensive theories of the housing market, where a clear conception of the systemic nature of residential activities is evident'.

The Ecological Approach

First developed by the Chicago School in the 1920s, the approach is based on biological analogy. Since man is an organic creature, it was argued that he is subject to the general laws of the organic world, most notably those of competition, dominance and invasion and succession. According to Park (1952, p. 119)

> competition, which is the fundamental organising principle in the plant and animal community, plays a scarcely less important role in the human community. In the plant and animal community it has tended to bring about (1) one orderly distribution of the population and (2) a differentiation of the species within the habitat. The same principles operate in the case of human population.

Thus man competes for space and access to the most desirable locations and the level of competition is reflected in land values. In this way, households are segmented into distinct spatial categories according to their ability to pay.

As with competition, 'the principle of dominance operates in the human as well as in the plant and animal communities' (Park, 1961, p. 25). In plant associations, one species exerts a dominant influence by controlling the environmental conditions which encourage or discourage other species. According to the Chicago School, the CBD is the area of dominance within the city but within each area or zone different classes or categories of activity are dominant and resist encroachment. Thus industry tends to repel residential development, and high income households resist the infiltration of their areas by low income families, etc.

Perhaps the most famous example of the ecological approach is the model developed by Burgess in 1924. Based on a study of Chicago, it identified five concentric zones (the CBD, the zone in transition, the zone of working men's houses; the zone of better residences and the commuters' zone) and suggested that over time a city would expand out from its centre in a series of concentric zones. Once again, the process by which a city grows and expands is seen to follow biological principles. Over time, plants change their micro environment and, frequently, create conditions in which other plants are able to thrive. These species invade the environment and eventually establish themselves as the dominant element. This is the process of invasion and succession and according to Burgess a city grows and expands as a result of the build up of pressure in the central area forcing population to invade the next outer area and eventually to succeed the host population.

Since its inception, the Burgess model has been the focus of considerable attention and has been criticised, tested, modified and extended. Most of this work is reviewed in Johnston (1971) and it is necessary here only to refer to the main issues. For instance, early criticism of the approach focused on the distinction between the biotic and cultural aspects of society (Alihan, 1938) and the relative neglect of cultural factors (Firey, 1945). In comparison, more recent criticism (Senior, 1973) has stressed that the urban ecological principles only suggest that different socio-economic groups will become segregated and that the concentric zonation of residential areas is just one of many possible spatial expressions. Indeed, Senior argues that the pattern identified by Burgess is a consequence of certain quite restrictive assumptions about the locational preferences of high status groups and the role of landowners and developers.

In a series of papers dating from the early 1960s, Schnore (1963, 1964, 1965, 1966) has attempted to test the validity of the Burgess model and he concludes from these detailed investigations that the model is valid only for the largest and oldest of the US cities. Indeed, it would seem that the model represents the final stage in the evolution of an urban system. In the pre-industrial city, the pattern appears to be reversed and with modernisation, it is the middle income groups which migrate to the suburbs.

Undoubtedly the most notable modification has been that proposed by Hoyt (1939). From a study of 142 North American cities, Hoyt concluded that residential areas tend to conform to a pattern of sectors rather than concentric circles (ibid., p. 76) and that the ultimate residential pattern is determined by the pattern of high status areas. These, he suggested, radiate out from the centre of the city along major

Table 1.2: Social Area Constructs

Constructs	Variables
Economic status (social rank)	Occupation, schooling, rent, single family dwellings.
Family status (urbanisation)	Fertility, female employment.
Ethnic status (segregation)	Racial and national groups.

Source: After Shevky and Bell (1955).

transportation routes which 'pull' high value property away from pre-existing industrial areas towards the more open countryside. Fundamental to this view is the belief that cities grow as a result of the periodic demand for new property by the highest income groups. As dwellings obsolesce, the most well-to-do demand new property (on the edge of the built up area) and their homes are occupied by lower income households (the filtering process).

Although this sector pattern has been found to be applicable in several areas (Jones, 1961; Robson, 1966) it has been argued that the traditional ecological approach 'has not been particularly illuminating owing to vagueness over the actual characteristics of residential structure for which Burgess was proposing his concentric model' (Senior, 1973, p. 173). Of more significance, perhaps, have been social area analysis and factorial ecology.

One of the features of the ecological approach is that the models proposed by Burgess and Hoyt are complementary and not contradictory. Indeed it should be appreciated that invasion and succession is a special case of filtering which can lead to the demand for new housing, while Hoyt was at pains to point out that distinct zones can be identified within sectors. It is not surprising to discover, therefore, that many cities display features characteristic of both models and studies using social area analysis have demonstrated that there may be more than one dimension of differentiation within cities (Anderson and Egeland, 1961). The method of social area analysis was first used by Shevky and Williams (1949) in their study of residential patterns in Los Angeles and by Bell (1953) in San Francisco. Basically, it involves the derivation of three constructs to summarise the differentiated characteristics of the residential population (Table 1.2). These three general indices, derived from selected census tract variables, are used to produce an 18 cell classification. Each census tract with similar values across the three indices is then allocated to one of the cells or social areas. Although the theoretical underpinning of social area analysis is weak and has been criticised on numerous occasions (Hawley and Duncan, 1957), tests by Bell (1953) and Van Ardsol *et al.* (1958) among others have tended to confirm the empirical validity of the constructs, though from the few studies conducted outside of America, (McElrath, 1962; Herbert, 1967; Abu-Lughod, 1969) it would seem to be less applicable. However, using data from four American cities, Anderson and Egeland showed that the distribution of socio-economic variables was sectoral, whilst life cycle variables were zonal. These findings have been substantiated by the results of studies in Australia (Timms, 1971), Canada (Murdie, 1969) and England (Robson, 1969). In studies such as these, multivariate statistics (factor analysis, principal components analysis or cluster analysis) were used to examine the spatial patterns of residential differentiation and to produce a factorial ecology of the city based on the indices proposed by Shevky and Bell. However, the factorial ecologies differ from the analyses of social areas in that 'whereas the Shevky technique selects its constructs, and the variables which compose them, on the basis of possibly suspect theory, multivariate analysis selects its discriminating factors solely on the basis of the intercorrelations of the data itself' (Robson, 1969, p. 58). In the 1960s and early 1970s such studies constituted a major research focus in urban geography not only in the cities of the advanced countries of Australasia, North America and Western Europe, but also in third world cities (Berry and Rees, 1969; Hill, 1973). In most instances, the results confirmed that the economic status indexes aligned sectorially (Figure 1.2A), the family status indexes aligned concentrically (Figure 1.2B) and the ethnic status indexes clustered in particular parts of the city (Figure 1.2C). Also, they tended to support the contention that classical ecological models were complementary rather than competitive and that it was possible to develop an integrated spatial model of the residential area of a city, as proposed for Chicago by Berry and Rees in 1969 (Figure 1.2D).

By the mid 1970s, the focus of urban geographical research had changed and relatively few ecological studies were being conducted. Increasingly it was recognised that the ecological approach was essentially descriptive and that while the findings which it generated suggested hypotheses of the residential locational behaviour of households, the ecological models did not incorporate these hypotheses. Thus, the approach provided a description of urban residential structure which could be 'used as a foundation on which to build models that attempt to incorporate "how" and "why" residential location patterns come about' (Senior, 1973, p. 177).

Figure 1.2: An Integrated Spatial Model of Residential Structure (after Berry and Rees, 1969)

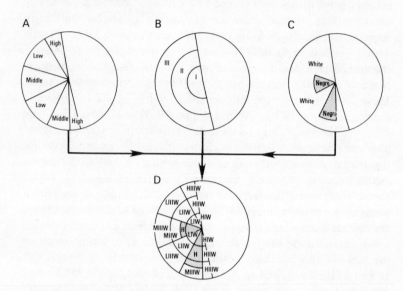

The Behavioural Approach

The behavioural approach to human geography was a direct response to the failure of the more traditional approaches to explain the spatial patterns which they had identified. Essentially, the approach focused on decision-making and in the field of housing, the main research foci have been the decision to move to a new dwelling and the search for, and choice of, a new home (Brown and Moore, 1970).

In most urban societies, the level of residential mobility (that is the movement of households within an urban area) is remarkably high and from the evidence available (Albig, 1933; Rossi, 1955) it would appear to be increasing. As Herbert (1972, p. 244) has observed

> All the available evidence would suggest that in general North American rates of residential mobility are much higher than in Europe. In Britain, for example, the average annual rate presently runs at between 8 and 10 per cent, but this figure shows signs of gradually increasing.

Clearly, this movement of people from one residence to another is 'the very mechanism by which zonal and natural areas . . . are created' (Rees, 1970, p. 307), especially since it would seem that mobility rates are not uniform throughout society. This has led to the identification of a mover/stayer dichotomy (Goldstein, 1954) which suggests that although movers come from all sectors of society (Whyte, 1960), more movement occurs within the rental market than within the owner occupier market and that the probability of a change of house is negatively related to the length of residence (Wilkinson and Merry, 1965; Morrison, 1967).

(a) Reasons for Moving

The vast majority of the studies have focused, inevitably, on the decisions of the movers, especially those moving out of choice and the results suggest that mobility takes place for two main reasons – either to improve or maintain a household's social position or to satisfy the changing needs of a household consequent on changes in the family life-cycle. Most evidence (Butler, Sabagh and van Ardsol, 1963; Maisel, 1966; Simmons, 1968) seem to support the latter explanation and in his seminal study of why people move, Rossi (1955, p. 9) suggests that residential mobility is 'the process by which families adjust their housing to the needs that are generated by shifts in family composition that accompany life-cycle changes'.

Precise definitions of the family life-cycle vary but Abu-Lughod and Foley (1960) have identified six distinct stages (pre-child, child-raising, child-rearing, child-launching, post-child, later-life). Although not mutually exclusive and not followed by every household, these can be related to the three life-style patterns identified by Bell (1958). The result is a series of moves at differing stages in the life-cycle to properties of different types and sizes in different parts of the city to satisfy the changing needs of the household (Table 1.3). However, although 'there is widespread professional acceptance and use of such a generalised scheme' (Foote *et al.*, 1960, p. 362), evidence produced by Leslie and Richardson (1961) from a study of a relatively new suburban area in Lafayette, Indiana, suggests that life-cycle variables are of little assistance when predicting residential mobility, which is best explained by social mobility expectations, perceived class differences, education levels and attitudes to present dwelling. From this, Leslie and Richardson argued for a much broader explanation of residential mobility than Rossi had proposed based on both social mobility and life-cycle changes. Support for this view has come from Brown and Moore (1970) who suggested that the decision to move is a product of the stress generated

Table 1.3: Combined Life-cycle and Life-style Model of Residential Preference

Stage	Style	Space	Median tenure	Housing age	Mobility	Locational preference
I Prechild	Consumerism/ Careerism	Unimportant	Rented flat		1 move to own home	Centre city
II Child-bearing		Increasingly important	Rented house	Old	High 2–3 moves	Middle and outer rings of centre city
III Child-rearing	Familism/ Careerism	Important	Owned	Relatively new	1 move to owned home	Periphery of city or suburbs
IV Child-launching		Very important	Owned	New	1 move to second home	Suburbs
V Postchild	Consumerism/ Careerism	Unimportant	Owned	New when first bought	Unlikely to move	
VI Later life	Familism	Unimportant		Widow leaves owned home to live with grown child		

Source: After Abu-Lughod and Foley (1960) and Bell (1958).

by the disparity between the needs and aspirations of a household and the qualities of its actual home and environment.

Although these studies have done much to aid the understanding of the decision to move, the majority have been conducted in America and are not necessarily typical of the situation elsewhere in the western world. Equally, they are not without their limitations. As Morgan (1976) discovered in Exeter, factors other than stages in the family life-cycle frequently intervene in the choice of a home, while Rainwater (1966) has shown that many working-class households view their homes not as a basis for the enactment of certain favoured life-styles, but rather as a haven from the outside world. What is more, most of the literature has focused on the residential moves of the higher and middle income households and, as mentioned above, upon those who move, especially those who move voluntarily. In so doing, the emphasis has been placed upon freedom of choice and little attention has been paid to the way choice is constrained. For instance, many households are unable to move because of the lack of resources or, as Kirby (1973) and Bird (1976) have demonstrated, because of the constraints of tenure. Alternatively, many households are forced to move through dramatic neighbourhood change (for example, slum-clearance) or family circumstances (for example, divorce). Clearly, in such situations, choice enters only in the decision about where and what to move to, and even then many of those households forcibly moved may have little, if any, say in the type or location of the property made available to them (Gray, 1976). These households apart, however, all relocating households must search for suitable vacancies and decide upon a new home. It is this procedure which has focused as the second major area of behavioural research.

(b) The Search for a New Dwelling and the Choice of a Home

When a household moves, it must determine the type and location of housing it desires and several studies (Troy, 1973; Herbert, 1973) have examined the criteria used to define this 'aspiration zone'. Generally, these criteria are found to be related to the reasons for moving and it would seem that the more important are cost, dwelling characteristics, location, quality of the physical environment and the social status of the neighbourhood. However, the relative importance of these factors is found to vary according to the socio-economic status of the household.

A second area of investigation has been the search procedure. Once the upper and lower limits of acceptability have been defined, the right dwelling has to be found in the time available, at the right price and in the correct location. One of the most consistent findings of research into residential mobility has been that the majority of moves are relatively short (Rossi 1955; Adams, 1969; Butler *et al.*, 1969; Johnston, 1969; Speare *et al.*, 1975). Clearly, one of the factors responsible for this is the size of the city. Another, however, is the nature of the search procedure. According to Adams (1969) and Brown and Moore (1970), households possess an area of knowledge (an 'awareness space') of the city in which they reside, based upon the location of their present residence. From the evidence available (Brown and Holmes, 1971; Barrett, 1973), it would seem that especially for low income households, the area of research is concentrated around the former residence and that most households relocate, in fact, within their awareness space.

This feature is reinforced by the fact that:

(1)   most moves are made to areas with a socio-economic status similar to the area of origin (Goldstein and Mayer, 1961; Brown and Longbrake, 1970; Speare *et al.*, 1975; Clark, 1976 and Short, 1978) and areas of similar socio-economic status do tend to cluster together within a city, as the preceding discussion has demonstrated;

(2)   the information sources which a household can use when searching for a new home exert spatial bias. This is a relatively under-researched area but Rossi (1955) has pointed to the high proportion of households in Philadelphia relying upon newspapers and personal contact, while Michelson (1977) in Toronto has stressed the importance of estate agents and Herbert (1973) in Swansea has observed that informal sources (personal contacts and 'looking around') are particularly important for the lower income groups. Each of these sources is spatially biased. Although newspapers, perhaps, carry the widest range of housing and possess the least spatial bias households tend to look for advertisements in areas known to contain dwellings of the desired type and price. Similarly 'looking around' is closely determined by the household's awareness space, while personal contacts are influenced by both social class and by propinquity (Gans, 1961). Both of these have a localising influence as can family ties, especially in low income areas. Finally, estate agents tend to specialise in property within certain price ranges and areas and thus households depending upon estate agents 'are making use of a

highly structured and spatially limited information source' (Palm, 1976, p. 28).

The third line of investigation has been an examination of the factors influencing the actual choice of a home and in a highly original and innovative paper, Flowerdew (1976) has outlined a set of stopping rule models based on the assumption that the prospective buyer decides to stop by weighing 'the possibility of finding a more suitable invest-ment against the cost and risks of further search' (MacQueen and Miller, 1960, quoted in Flowerdew, 1976, p. 50). Given the variability of the decision-making process, however, the work does not appear to have been developed further and research into the actual decision about the choice of a home is still not highly advanced. However, it would seem (Barrett, 1973) that in an attempt to reduce uncertainty, households consider only a small number of vacancies over a limited area in a short time period, while Lyon and Wood (1977, p. 1175) conclude from their study of the house purchase decision on a modern estate in South Croydon, that because of the constraints of time, information and motivation, 'the choices made by people are made in a far less rational manner than one might have thought possible'. Indeed, this is one of the weaknesses of the behavioural approach, as mentioned earlier — it fails to consider the constraints which are placed on demand. The more recent approaches have tended, therefore, to concentrate on these con-straints, particularly in terms of the constraints imposed by the agencies responsible for the supply and allocation of housing. Before considering these studies, however, it should be recognised that over time changes occur in the patterns of residential areas within cities and although mobility need not lead, necessarily, to residential change, 'changes in the residential pattern of a city cannot occur independent of migration' (Johnston, 1971, p. 296). Accordingly, several studies have examined the mechanisms of change concentrating, in particular, on vacancy chains and the filtering process (Lansing *et al.*, 1969; Guy and Nourse, 1970; Watson, 1973 and Dzus and Romsa, 1977) and the reverse process of gentrification. Very simply, this is the 'upgrading' of inner city areas through the 'invasion of traditionally working class areas by middle and upper income groups' (Hamnett, 1973, p. 252). Most of the large cities in the western world now display examples of such areas and apart from the early work by Hamnett, investigations have been under-taken by Cybriwsky (1978) in Philadelphia, Gale (1979) in Washington and Hamnett and Williams (1980) in London, among others.

Supply-based Explanations

In the traditional demand-based explanations of housing, decisions were seen as the result of the freedom of choice; individuals and households were believed to be free to choose where, and in what type of housing, they wanted to live. In reality, this is not the case and in a criticism of the behavioural approach to explanation in urban geography, Gray (1975, p. 230) has observed that

> people are not free to choose and prefer from a range of options and that the study of household units does not provide the key to understanding urban processes. Instead, many groups are constricted and constrained from choice and pushed into particular housing situations because of their position in the housing market, and by the individual institutions (i.e. building societies, estate agents, public and private landlords) controlling the operation of particular housing systems.

What is more, even where freedom of choice does appear to exist, frequently the range of choices available has been predetermined by the decisions and actions of the agencies responsible for the supply and allocation of resources. Thus it has been argued that 'the housing market process can only be understood as part of the national allocation of resources, and not as the outcome of demand from individual household units' (Mellor, 1973, p. 39). Accordingly, the more recent studies have focused on the factors affecting the supply of housing services.

The Managerialist Approach

The basis for much of this work has been Weberian sociology which sees man as an 'actor' fulfilling a variety of roles which, quite frequently, conflict. As Lambert *et al.* (1978, p. 6) have observed, it describes a society 'where interest groups collide, collude and cohere in the control of institutions, where privilege and status are negotiated, where, in short, power becomes the crucial variable'. In the field of housing, man can be regarded as fulfilling two broad roles, those of housing occupier and those of housing supplier and in a study of Sparkbrook in Birmingham, Rex and Moore (1967) focused on man as an 'occupier'. Following the Weberian approach, they suggested that households could be grouped into distinct housing classes based on their degree of access to scarce housing opportunities and argue that not only do these classes have a distinct spatial distribution within the city, but that the mechanism behind the housing market is 'a class struggle between groups differentially placed with regard to the means of housing' (Rex, 1968, p. 215).

Since its publication in 1967, the work of Rex and Moore has been severely criticised on several counts, most notably by Pahl (1970) who, despite all of the criticisms, used 'Rex and Moore's discussion of housing classes and locality-based conflict groups' (ibid., p. 247) to develop a 'new urban sociology' based on the organisation of resources and facilities. He argued that 'the controllers, be they planners or social workers, architects or education officers, estate agents or property developers, representing the market or the plan, private enterprise or the State, all impose their goals and values on the lower participants in the urban system' (ibid., pp. 207-8) and 'we need to know how the basic decisions affecting life chances in urban areas are made' (ibid., p. 208). This has become the basis for the managerialist approach.

Since the mid-1970s, managerialism has provided an important focus for housing research, particularly in Britain where 'welfare capitalism, with its institutions of social and economic policy modifying the free play of market forces, has produced an easily identifiable bureaucratic layer in the British housing scene' (Bassett and Short, 1980, p. 51). Essentially, the studies have concentrated on the 'gatekeeper' role of the various institutions controlling entry into the various housing sub-markets. These institutions, identified by Harvey (1973) as estate agents (realtors), landlords, developers, financial institutions and government institutions, operate in such a way that they 'cannot easily be brought together into one comprehensive framework for analysis' (ibid., p. 166). Accordingly, the various studies of housing management have tended to be empirical examinations of the individual institutions.

(i) Estate Agents. Reference has been made already to the role of estate agents in the choice of a home and several studies (Rossi, 1955; Barrett, 1973; Herbert, 1973 and Short, 1977) have examined the way the different information channels are used by prospective buyers of housing. From such studies, it would appear that the higher income groups are most likely to rely on estate agents and other formal information channels, while the lower income households are more likely to use informal sources of information such as personal contact. However, estate agents are not simply passive suppliers of information. Rather, they play an active part in the house-purchase decision by deliberately channelling the prospective purchaser into, or away from, specific areas. This social gatekeeping role has been examined by

Williams (1976a) and Palm (1979) among others, while attention has also been paid to the way estate agents tend to steer coloured buyers away from certain areas in both British (Burney, 1967) and American (Helper, 1969; Brown, 1972) cities.

While estate agents can, through their gatekeeping role, help to preserve the social status of an area, they can equally contribute to urban change. As Williams (1976b) has demonstrated, estate agents 'through their role as advisers to landlords, building societies and property companies' (ibid., p. 80) can stimulate the gentrification process by persuading the financial institutions to lend money for the purchase and rehabilitation of obsolete property. Frequently, however, they also bring about change by purchasing property themselves and extending their role to that of the landlord developer and/or speculator. When this occurs, the end product can be change in the form of either gentrification or residential deterioration. Usually the latter involves the deliberate sale of houses in predominantly white areas to coloured families in an attempt to stimulate the migration of white households to other localities, though in some instances estate agents have been found to purchase property and to deliberately allow it to deteriorate, thereby initiating out-migration from the area and enabling more of the property to be purchased and the land to be sold, eventually, as a prime site for redevelopment.

(ii) Landlords. In the traditional explanations attention was focused predominantly on the decisions and behaviour of the home owner. Much of the literature and research emanated from North America and Australia where home ownership levels are in the order of 60-70 per cent. Elsewhere in the western world, however, the level of home ownership is not so high. In Germany and Sweden, for instance, only about one-third of the population own their own home, though admittedly the proportion is increasing.

With the advent of the managerialist approach to housing, there has been increasing awareness of, and interest in, the rental sector and studies have focused on the role of both the private and the public landlord. In countries like America, Australia and Canada, public housing accounts for only a small proportion of the dwelling stock and private renting constitutes the main, if declining, alternative to home ownership. In Britain and western Europe, however, public sector housing constitutes a considerably higher proportion of the total dwelling stock (though normally not more than 40 per cent) and since the early twentieth century has been seen increasingly as an integral feature of the welfare systems of these countries. In contrast, private rental has tended to decline in significance and in Britain it now accounts for no more than 16 per cent of the total stock. Even so, both private and public sector rental constitute an important feature of the housing systems of most western-style economies and in recent years the procedures by which rented accommodation is allocated have become an important focus for housing research.

(a) Private Landlords. Within the private sector, several different types of landlord can be found. In Edinburgh, Elliot and McCrone (1975) have identified individuals, property companies, trusts, welfare organisations, commercial firms which have diversified into property and public institutions (such as churches and universities). By far the majority are small, individual landlords as Cullingworth (1963) and Greve (1965) have demonstrated but of most importance, in terms of their control over the market, are the large organisations and property companies. While relatively little is known about the attitudes and behaviour of these various groups, it is generally held that the property companies can be viewed as profit maximisers, allocating property on a strict 'ability-to-pay' basis in a formal bureaucratic manner. As Bassett and Short (1980, pp. 84–5) have observed 'large landlords and property companies often adopt allocation rules that select higher-income, childless, professional households who can pay higher rents and cause fewer management problems'. In contrast, individual landlords often allocate their properties less bureaucratically and not necessarily on economic grounds, taking into consideration personal factors and subjective 'character' assessments. A similar procedure has also been found among certain institutional landlords where the dominant criterion frequently has been social need rather than the ability to pay (Harloe *et al.*, 1974). However, it would seem that because of the domination of demand over supply (and certain aspects of housing legislation) all private landlords are able to select tenants and to discriminate against unsuitable or undesirable households by refusing them access to rented accommodation. Generally these households appear to be the noisy and unreliable tenants, long-stay tenants and, increasingly, households with children (Short, 1979). Thus although there are no formal qualifications required to gain access to privately rented housing, the gatekeeping activities of private landlords inevitably influence who gets what and goes where.

(b) Public Landlords. By contrast, formal qualifications are required to gain access to council housing and considerable attention has been paid, in recent years, to the role of the local housing authority.

Basically research has focused on two separate, but related aspects, namely the criteria adopted for determining a household's eligibility for housing and the way in which dwellings are allocated to households. As English (1979, p. 117) has observed, the former function was, until fairly recently, of paramount importance but 'the second function of housing allocation, deciding who shall have which house, has become the more important'.

Essentially the allocation of public housing is based on the principle of housing need. Although there are no strict, uniform criteria, it is generally held that a local authority has responsibility for housing the homeless, those requiring rehousing because of slum clearance or redevelopment, those in urgent need of housing because they are regarded as 'key' workers or have acute health or social problems (such as overcrowding), those already living in public housing who wish to transfer to alternative accommodation and those who are not in need but wish to be considered for council housing. As in the private sector, however, the demand for public housing invariably exceeds supply and most local authorities operate a waiting list. These lists vary, as Niner (1975) has shown from simple queuing systems based on the length of time on the list and the availability of suitable accommodation to sophisticated 'points' schemes based on the criteria outlined above.

Since council housing is intended to meet the housing needs of the district, it is inevitable that variations in policy exist and owing to the differing weights placed by the various housing authorities on the different aspects of housing need, it is not unusual for households with identical circumstances, in different local authority areas, to have quite different degrees of access to council housing (Murie, Niner and Watson, 1976). Even so, although there is variation, it is variation in details and not in fundamentals and, as Niner (1975, p. 44) discovered, there is 'a general consensus on the purpose of council housing, who should be eligible for tenancy and the elements to be taken into account in considering and defining housing need'. Generally those least likely to gain access to state housing are the young, single households without dependants, newcomers and former owner-occupiers, though it is also possible that those in greatest hardship are missed 'because definitions of need in operation are not those most appropriate' (ibid., p. 94).

The second major focus has been the allocation of housing to tenants and several studies have investigated the screening procedures adopted by local authorities and the role of the housing investigator (Damer and Madigan, 1974; Gray, 1976). Ostensibly the purpose of screening is to ensure that prospective tenants are allocated the dwellings most suited to their needs. In reality what this means is that the poorest households are allocated the poorest housing. While local authorities are reluctant to admit to such a procedure, housing authorities in Britain were advised to adopt such a policy (Central Housing Advisory Committee, 1956) though they were warned not 'to rehouse too many problem families in one street' (ibid., p. 13). By the late 1960s, however, the criteria for access to new housing seemed not to be housing need but 'moral rectitude, social conformity, clean living and a "clean" rent book' (Central Housing Advisory Committee, 1969, para. 96). What is more, many local authorities were concentrating problem families in the least desirable estates. Under such circumstances a frequent feature of the dwelling stock in many urban areas was the problem council estate. Numerous studies have identified such estates and examined their problems and characteristics (Wilson, 1963; Baldwin, 1974; Griffiths, 1975; Attenburrow *et al.*, 1978), noting, in particular, the role of the local authority in their creation, the stigma associated with living on such estates and the difficulties experienced by tenants in transferring from them and by local authorities in filling them (Taylor, 1978). In several areas dwellings on these estates are now being occupied by households which normally would be ineligible for public housing but who are desperate for accommodation, while Parker and Dugmore (1976) have demonstrated that in Greater London, the older, poorer-quality estates are being used increasingly to house coloured tenants. Thus, as Gray (1976, p. 42) has observed 'both the life chances of various groups in the city and the socio-spatial structure of urban areas are influenced by the selection and allocation procedures employed by local authority managers'.

(iii) Developers and Builders. In most large cities, the demand for housing exceeds supply and under such circumstances consumer preferences concerning the location and type of new building are relatively unimportant. As Craven and Pahl (1975, p. 112) observe what is important 'is the conception of such preferences in the builders' mind. Market conditions permit him to lead and educate rather than to follow demand'. It would seem, therefore, that within the limits of the planning framework, it is the decisions and behaviour of the builders and developers (that is the suppliers of housing services) that control the contemporary pattern of residential development within cities, rather than the decisions and behaviour of the

occupying households. However, apart from work by Chamberlain (1972). Harloe *et al.* (1974), Craven (1975) and Bather (1976) surprisingly little is known about the way such decisions are made and 'there is a basic lack of knowledge about these people and the way they interact' (Craven and Pahl, 1975, p. 110).

What these studies have revealed, however, is that the building industry tends to be characterised both by a small number of very large firms and by a large number of very small firms. Although the objectives of both groups are the same, to make profit, the methods by which this is achieved vary considerably. For instance, it would seem that the operations of the small-scale traditional builder tend to concentrate on the development of expensive detached housing for the higher income groups on small infill sites. In contrast, the large-scale builder, which has come to dominate the post-war construction industry, has tended to favour the construction of mass-designed housing on large sites in the urban periphery. According to Checkoway (1980, p. 22), it was the ability of the 'large builders to take raw surburban land, divide it into parcels and streets, install needed services, apply mass production methods to residential construction and sell the finished product to unprecedented numbers of consumers' which resulted in the growth of the American suburbs. In Britain and much of western Europe, suburbanisation has not developed on the same scale as in America but, even so, many European cities possess vast homogeneous suburban housing developments, reflecting the activities of the large, non-local builder. Usually these estates are characterised by uniform semi-detached or terraced housing intended for the lower and middle income groups, particularly 'young households in the early stages of the life-cycle, manual, semi-skilled and white-collar workers, immigrants from the metropolis and those moving from a lower-priced housing market in other parts of the country' (Craven, 1975, p. 138). Thus the social, economic and demographic characteristics of an area can be seen as the product of the decisions and activities of private builders and developers regarding the size of the development and the type of housing constructed.

The same is true in the public sector where local authorities act both as housing managers and as developers. Thus governments can determine not only where households live, but also the conditions under which they live. Several geographical studies have examined the role of the local housing authority as developer (Thomas, 1966; Blowers, 1970; Jennings, 1971; Kirby, 1971 and 1974; Kivell, 1975; Paris and Lambert, 1979) and while the pattern of development appears to be very similar to that of the large private developer, there is more variation with local authorities building a broader range of housing types (including high-rise flats, aged persons' dwellings and sheltered housing) on a wider variety of sites and in different urban locations, including the inner city. In this way, local authority development has greatly influenced the residential structure of cities and modified considerably the patterns proposed by the traditional demand-based approaches.

(iv) Financial Institutions. According to Stone (1978, p. 190) the allocation of finance 'exerts a decisive influence over who lives where, how much new housing gets built and whether neighbourhoods survive'. Several studies have examined, therefore, the influence of the main financial institutions — the building societies, local authorities, banks and insurance companies — on residential structure. While much of this material relates to the British case, it would seem to be relevant elsewhere as the basis of the decision to lend appears to be the security of the loan. In reaching their decision the more reputable operations take into consideration both the credit worthiness of the applicant and the durability and sales potential of the property. Both of these criteria are biased against certain classes of household and certain types of property as the studies by Barbalet (1969), Boddy (1976), Duncan (1976), Lambert (1976) and Williams (1976b) have revealed. Essentially lending policies are found to be biased against low-income households and households headed by manual workers, as well as against older property and property in multiple occupation. Such policies 'produce a distinctive spatial pattern to the flow of funds within the city' (Bassett and Short, 1980, p. 76) by directing funds away from the inner city to housing on the urban fringe. As the Chief General Manager of the Nationwide Building Society has observed, building societies 'are reluctant to lend on properties which may be acceptable in themselves, if the environment in which the houses are situated is declining and the values of the properties are, therefore, likely to depreciate' (quoted in Williams, 1978, p. 25). This procedure of designating whole areas as being unsuitable for loan security is known as redlining and has been recognised by Vitarello, 1975; Boddy, 1976; Lambert, 1976; Weir, 1976 and Williams, 1978. Not infrequently such areas are associated with high concentrations of coloured households and it would seem that the penetration of an area by such households is perceived by the lending agencies as further evidence of neighbourhood instability and declining property values (Duncan, 1976; Weir,

1976). According to Boddy (1976) this denial of funds to the inner city by the lending agencies, but particularly the building societies, encourages decay and stimulates inner city decline. Frequently, local authorities advance loans to households for property considered unsuitable by the building societies. However, with the reduction in public spending experienced since the middle of the 1970s many marginal borrowers have been deprived of local authority funds and have had to rely on alternative sources. These have ranged from the major banks to what Duncan (1976, p. 314) has termed '"back-street" finance companies and moneylenders'. Usually such loans are short-term and interest rates are high. To meet the high interest rates 'some owners will take in lodgers to help meet the repayments and/or will have little money to spend on house repairs' (Bassett and Short, 1980, p. 81). The resultant overcrowding and neglect of maintenance results, inevitably, in the further decline of inner city housing.

In certain circumstances, however, the policies of the lending societies can lead to the improvement of inner city areas as Williams (1976b) has demonstrated in Islington. Here, the increased demand for property from the middle classes (who were often building society investors) resulted in 'an improving environment and an increasingly buoyant local housing market' (ibid., p. 78). By the late 1960s, there-fore, successful applicants for mortgages were no longer exceptional though 'the beneficiaries of this change in attitude were the affluent newcomers . . . rather than the local populace, because even though finance became steadily easier, terms were still stringent and sub-stantial amounts of cash were necessary for both downpayment and improvement' (ibid., p. 78). As a consequence, the area has been gentrified and it would seem that in addition to consumer choice 'factors of supply, of control and of constraint are of equal if not greater importance' (ibid., p. 72) in determining the residential structure of urban areas.

(v) Central and Local Government. As the preceeding discussion has indicated, government intervention in the housing market considerably influences the ultimate residential structure of cities. However, the activities of government are not confined to the roles of housing developer and manager. In Figure 1.3, the main forms of government intervention are outlined and it can be seen that policies relating to both public and private sector housing are formulated at the national level and implemented through the local government system. Several studies have examined the nature and development of public housing

Figure 1.3: Public Intervention in the Housing System

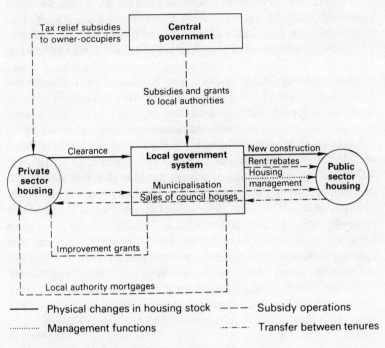

Source: Bassett and Short (1980).

policies (Wendt, 1963; Donnison, 1967; Wolman, 1975; Headey, 1978; Stafford, 1978; Lansley, 1979; Balchin, 1981 and McGuire, 1981) and the reaction of the local housing and planning authorities to these policies (Murie, 1975). For instance, Murie *et al.* (1976) have observed that while government policy does reflect political ideology, it is affected, also, by the economic climate and in Britain appears to have 'developed in a haphazard fashion and in response to particular crises' (ibid., pp. 241-2). Similarly, they point out that there is a 'wide variation in the way in which housing policy is implemented in different parts of the country' (ibid., p. 241) and this can, and does, have a fundamental effect on access to, and choice in, housing.

Most studies have focused on government policies towards residential renewal (Kirby, 1979) and much of the early research in this area concentrated on slum clearance. In particular, attention was paid to the failure of the clearance programmes to resolve the problems of slum housing and to the social problems created by planned rehousing

(Hole, 1959; Montgomery, 1960; Morris, 1962; Gruen, 1963; Hartmann, 1964; Gans, 1965; Friedman, 1967). In contrast, later work was more concerned with the relative merits (both social and economic) of rehabilitation and clearance (Montero, 1968; Bagby, 1973; Listokin, 1973) and in an influential consideration of the economics of housing, Needleman (1965) initiated a detailed discussion of the rehabilitation/redevelopment decision (Sigsworth and Wilkinson, 1967; Needleman, 1968; Schaaf, 1969). More recent studies have followed a politico-economic approach and in Britain have focused on such issues as the management of clearance and redevelopment (English *et al.*, 1976) the take-up of government grants for the improvement/rehabilitation of older housing (Kirwan and Martin, 1972; Duncan, 1973; Duncan, 1974) and the achievements of policies aimed at directing aid to areas of greatest housing need, mainly in the inner city (Bassett and Short, 1978; Balchin, 1979). What these studies revealed is that the grants for improvement have tended to benefit most the better-off owner-occupiers living in localities peripheral to the poorest housing areas and that neither the general improvement area programme (Roberts, 1976) nor the housing action area programme (Paris, 1977; Short and Bassett, 1978) has modified substantially this pattern. Indeed, in many areas, the urban poor have been displaced as gentrification has occurred and property values have risen (Pearson and Henney, 1972; Hamnett, 1973).

Conclusion

As a result of the research undertaken throughout the latter part of the 1970s, there has emerged 'a much clearer picture of the political nature of housing, of the redistributional effects of the housing system and of the limits to the incrementalist managerial solutions which have tried to improve the workings of the system by smoothing the bottlenecks of allocation or by softening the impact of financial constraints in the production of houses' (Robson, 1979, p. 74). Even so, there is still 'much we need to know about the mechanisms by which housing is distributed in our society and about the changing conditions and occupancy of that housing' (Bourne, 1981, p. 262). As Hole (1979, p. 38) has emphasised 'the problems of housing remain substantial — the field is complex and there is only a limited understanding of the causal relationships and the processes operating within the system'. Urban managerialism, as applied to housing, has advanced considerably the body of knowledge on how, why and to whom scarce resources are allocated and has laid a sound empirical foundation for further research. However, it must be recognised that managerialism is merely a framework of study; it is not a theory of housing allocation and it seems inevitable that if further progress is to be made, it will be in 'the development of some more robust theoretical frameworks that one of the future foci of activity in the whole field of housing will lie' (Robson, 1979, p. 75). Similarly, it must be recognised that while the decisions of individual managers can affect the distribution of resources, it is unrealistic to assume that decisions are made in isolation and are independent of politico-economic influences. Accordingly, it seems appropriate that the study of urban managerialism should be related to a more general conception of the political economy of urban areas and to a more general theory of the city in capitalist societies. Such a view is central to the Marxist approach to housing, as epitomised by the writings of Harvey, 1973 and Castells, 1977. While the development of such an approach is not easy its potential lies in the fact that it focuses naturally on 'questions of power and conflict and the relationships between individuals and groups and on the processes of economic development' (Williams, 1978, p. 240). Accordingly, there seems little doubt that Marxist theories in relation to housing 'will receive even more attention from writers in various "schools" in the future' (Bassett and Short, 1980, p. 228).

While there is an obvious need for theory, there is also an immediate need for further empirical research. To date, much housing research has taken the form of systematic analyses of particular aspects of the housing system or of the roles of individual actors in the system. No attempt has been made to weight these influences and there is a need, now, for more synthetic studies of either individual urban areas or areas within cities which will attempt to determine the influence of the various factors within the housing system. In particular, attention needs to be focused on the relationship between household choice and the constraints imposed by the agencies of supply, thus unifying the managerial and behavioural approaches. Equally more attention needs to be paid to the identification of sub-markets and to the needs and requirements of minority groups, most notably the elderly, the infirm, the deprived, etc. As for any other product, demand for housing is not homogeneous and distinct consumer segments exist. It is necessary to know much more about the characteristics of these housing sub-markets and to know more precisely whether their needs are being met or their aspirations are being constrained. There is considerable scope, also, for further research into the effect of public policy on the housing system, particularly into the relationship between central and local government and between the housing system and non-housing policies. In this latter context, it is noticeable that there has been relatively little research into the impact of planning legislation on the housing system, despite the obvious importance of planning policy in guiding the pattern of residential development. Spatial variations in house prices also constitute a further area where research has been neglected. In this context, future research might be particularly fruitful if related to studies of the construction industry and the tightness of local markets.

From this list of potential avenues for research and the points made in the text, it is clear that there is considerable scope for further research into urban housing. Since the early 1970s, housing research has developed into one of the major foci of research into urban geography. While this situation is likely to continue as the subject develops both theoretically and empirically, it seems unlikely that any one focus will emerge. Indeed, if the purpose of research is to change and improve reality as well as to understand it, then perhaps 'priorities for research must reflect the problems facing particular local communities' (Bourne, 1981, p. 263).

References

Abu-Lughod, J. (1969) 'Testing the Theory of Social Area Analysis; the Ecology of Cairo, Egypt', *American Sociological Review, 34*, 189–212

Abu-Lughod, J. and Foley, M.M. (1960) 'Consumer Differences' in Foote, N.N., Abu-Lughod, J., Foley, M.M. and Winnick, L. (eds.) *Housing Choices and Constraints*, McGraw Hill Book Co., New York

Adams, J.S. (1969) 'Directional Bias in Intra-urban Migration', *Economic Geography, 45*, 302–23

Albig, W. (1933) 'The Mobility of Urban Population', *Social Forces, 11*, 351–67

Alihan, M.M. (1938) *Social Ecology: A critical analysis*, Columbia University Press, New York

Alonso, W. (1960) 'A Theory of the Urban Land Market', *Papers and Proceedings of the Regional Science Association, 6*, 149–58

Anderson, T.R. and Egeland, J. (1961) 'Spatial Aspects of Social Area Analysis', *American Sociological Review, 26*, 392–9

Attenburrow, J.J., Murphy, A.R. and Simms, A.G. (1978) *The Problems of some Large Local Authority Estates – An Exploratory Study*, Current Paper 18/78, Building Research Establishment, Garston, Watford

Bagby, D.G. (1973) *Housing Rehabilitation Costs*, Lexington Books, Lexington

Balchin, P.N. (1979) *Housing Improvement and Social Inequality*, Saxon House, Farnborough

Balchin, P.N. (1981) *Housing Policy and Housing Needs*, Macmillan, London

Baldwin, J. (1974) 'Problem Housing Estates – Perceptions of Tenants, City Officials and Criminologists', *Social and Economic Administration*, 8

Ball, M. and Kirwan, R. (1975) 'The Economies of an Urban Housing Market; Bristol Area Study', *Research Paper*, 15, Centre for Environmental Studies, London

Barbalet, R.H. (1969) *Housing Classes and the Socio-ecological System*, University Working Paper 4, Centre for Environmental Studies, London

Barrett, F.A. (1973) *Residential Search Behaviour: A Study of Intra-urban Relocation in Toronto*, Geographical Monographs no. 1, York University, Toronto

Bassett, K. and Short, J.R. (1978) 'Housing Improvement in the Inner City: A Case Study of Changes Before and After the 1974 Housing Act', *Urban Studies*, *15*, 333–42

Bassett, K. and Short, J. (1980) *Housing and Residential Structure: Alternative Approaches*, Routledge and Kegan Paul, London

Bather, N.J. (1976) *The Speculative Residential Developer and Urban Growth*, Paper 47, Department of Geography, University of Reading

Bell, W. (1953) 'The Social Areas of the San Francisco Bay Region', *American Sociological Review*, *18*, 39–47

Bell, W. (1958) 'Social Choice, Life Styles and Suburban Residence' in Dobriner, W. (ed.) *The Suburban Community*, Putman, New York

Berry, B.J.L. and Rees, P.H. (1969) 'The Factorial Ecology of Calcutta, *American Journal of Sociology*, *74*, 447–91

Bird, H. (1976) 'Residential Mobility and Preference Patterns in the Public Sector of the Housing Market', *Transactions, Institute of British Geographers*, *1*, 20–33

Blowers, A.T. (1970) 'Council Housing: The Social Implications of Layout and Design in an Urban Fringe Estate', *Town Planning Review*, *41*, 80–92

Boddy, M. (1976) 'The Structure of Mortgage Finance: Building Societies and the British Social Formation', *Transactions, Institute of British Geographers*, *1*, 58–71

Bourne, L.S. (1981) *The Geography of Housing*, Arnold, London

Brown, L.A. and Holmes, J. (1971) 'Search Behaviour in an Intra-urban Migration Context: A Spatial Prospective', *Environment and Planning*, *3*, 307–26

Brown, L.A. and Longbrake, D.B. (1970) 'Migration Flows in Intra-urban Space: Place Utility Considerations', *Annals, Association of American Geographers*, *60*, 368–87

Brown, L.A. and Moore, E.G. (1970) 'The Intra-urban Migration Process: A Perspective', *Geografiska Annalea*, *523*, 1–13

Brown, W.H. Jr. (1972) 'Access to Housing: the Role of the Real Estate Industry', *Economic Geography*, *48*, 66–78

Burgess, E.W. (1924) 'The Growth of the City: An Introduction to a Research Project', *Proceedings and Papers of the American Sociology Society*, *18*, 85–97

Burney, E. (1967) *Housing on trial*, Oxford University Press, Oxford

Butler, E.W., Sabagh, G. and Van Ardsol Jr M.D. (1963) 'Demographic and Social Psychological Factors in Residential Mobility', *Sociology and Social Research*, *48*, 139–54

Butler, E.W., Chapin, F.S., Hemmens, G.C., Kaiser, E.J., Stegman, M.A. and Weiss, S.F. (1969) *Moving Behaviour and Residential Choice: A National Survey*, National Co-operative Highway Research Programs, Report No. 81, Highway Research Board, Washington DC

Castells, M. (1977) *The Urban Question: A Marxist approach*, Arnold, London

Central Housing Advisory Committee (1956) *Moving from the Slums*, HMSO, London
Central Housing Advisory Committee (1969) *Council Housing: Purposes, Procedures and Priorities*, HMSO, London
Chamberlain, N.A. (1972) *Aspects of Developer Behaviour in the Land Development Process*, Research Paper 56, Centre for Urban and Community Studies, University of Toronto
Checkoway, B. (1980) 'Large Builders, Federal Housing Programmes and Postwar Suburbanisation', *International Journal of Urban and Regional Research*, *4*, 21–45
Clark, W.A.V. (1976) 'Migration in Milwaukee', *Economic Geography*, *52*, 48–60
Conzen, M.R.G. (1960) 'Alnwick, Northumberland: A Study in Town Plan Analysis', *Institute of British Geographers, Special Publication No. 27*
Craven, E.A. (1975) 'Private Residential Expansion in Kent, 1956–64: A Study of Pattern and Process in Urban Growth' in Pahl, R.E. *Whose City?* Penguin, Harmondsworth
Craven, E.A. and Pahl, R.E. (1975) 'Residential Expansion: A Preliminary Assessment of the Role of the Private Developer in the Southeast' in Pahl, R.E. *Whose City?* Penguin, Harmondsworth
Cullingworth, J.B. (1963) *Housing in Transition*, Heinemann, London
Cybriwsky, R.A. (1978) 'Social Aspects of Neighbourhood Change', *Annals of the Association of American Geographers*, *68*, 17–33
Damer, S. and Madigan, R. (1974) 'The Housing Investigator', *New Society*, *29*, 226–7
De Leeuw, F. (1972) *The Distribution of Housing Services*, Urban Institute, Washington
Donnison, D.V. (1967) *The Government of Housing*, Penguin Books, Harmondsworth
Drakakis-Smith, D.W. (1971) 'Slum Clearance at the Regional Level. The Establishment of Priorities for Renewal in the Context of Wales', *Town Planning Review*, *42*, 293–306
Duncan, S.S. (1974) 'Cosmetic Planning or Social Engineering?', *Area*, *6*, 259–70
Duncan, S.S. (1976) 'Self-help: The Allocation of Mortgages and the Formation of Housing Sub-markets', *Area*, *8*, 307–16
Duncan, T.L.C. (1973) *Housing Improvement Policies in England and Wales*, Research Memorandum 28, Centre for Urban and Regional Studies University of Birmingham
Dzus, R. and Romsa, G. (1977) 'Housing Construction Vacancy Chains and Residential Mobility in Windsor', *Canadian Geographer*, *21*, 223–36
Elliot, B. and McCrone, D. (1975) 'Landlords in Edinburgh: some Preliminary Findings', *Sociological Review*, *23*, 539–62
English, J. (1979) 'Access and Deprivation in Local Authority Housing' in Jones, C. (ed.) *Urban Deprivation and the Inner City*, Croom Helm, London
English, J., Madigan, R. and Norman, P. (1976) *Slum Clearance: the Social and Administrative Context in England and Wales*, Croom Helm, London
Evans, A.W. (1973) *The Economics of Residential Location*, Macmillan, London
Firey, W. (1945) 'Sentiment and Symbolism as Ecological Variables', *American Sociological Review*, *10*, 140–8
Flowerdew, R. (1976) 'Search Strategies and Stopping Rules in Residential Mobility', *Transactions, Institute of British Geographers*, *1*, 47–57
Foote, N.N. (1960) Abu-Lughod, J., Foley, M.M. and Winnick, L. (eds.) *Housing Choices and Constraints*, McGraw Hill Book Co., New York
Friedman, L.M. (1967) 'Government and Slum Housing: Some General Considerations', *Law and Contemporary Problems*, Spring, pp. 357–70

Fuson, R.H. (1964) 'House Types of Central Panama', *Annals of the Association of American Geographers*, *54*, 190–208

Gale, D.E. (1979) 'Middle-Class Resettlement in Older Urban Neighbourhoods', *Journal of the American Planning Association*, *45*, 293–304

Gans, H.J. (1961) 'Planning and Social Life, Friendship and Neighbour Relations in Suburban Communities', *Journal of the American Institute of Planners*, *28*, 103–40

Gans, H.J. (1965) 'The Failure of Urban Renewal: A Critique and some Proposals', *Commentary*, *39*, 29–37

Garrison, W.L. (1962) 'Comments on Urban Morphology' in Norbeg, K. (ed.) *IGU Symposium in Urban Geography*, Gleerups

Getis, A. (1969) 'Residential Location and the Journey from Work', *Proceedings Association of American Geographers*, *1*, 55–9

Goldstein, S. (1954) 'Repeated Migration as a Factor in High Mobility Rates', *American Sociological Review*, *19*, 536–41

Goldstein, S. and Mayer, K.B. (1961) *Metropolitanisation and Population Change in Rhode Island*, Rhode Island Development Council, Rhode Island

Gray, F. (1975) 'Non-explanation in Urban Geography', *Area*, *7*, 228–35

Gray, F. (1976) 'Selection and Allocation in Council Housing', *Transactions, Institute of British Geographers*, *1*, 34–46

Greve, J. (1965) *Private Landlords in England*, Bell, London

Griffiths, P. (1975) *Homes fit for Heroes: A Shelter Report on Council Housing*, Shelter, London

Grimshaw, P.N., Shepherd, M.J. and Willmott, A.J. (1970) 'An Application of Cluster Analysis by Computer to the Study of Urban Morphology', *Transactions, Institute of British Geographers*, *51*, 143–61

Gruen, C. (1963) 'Urban Renewals' Role in the Genesis of Tomorrow's Slums', *Land Economics*, August, pp. 285–91

Guy, D.C. and Nourse, H.C. (1970) 'The Filtering Process: the Webster Groves and Kankakee Cases', *Papers and Proceedings of the American Real Estate and Urban Economics Association*, *5*, 33–49

Hamnett, C. (1973) 'Improvement Grants as an Indicator of Gentrification in Inner London', *Area*, *5*, 252–61

Hamnett, C. and Williams, P.R. (1980) 'Social Change in London: A Study of Gentrification', *Urban Affairs Quarterly*, *15*, 469–87

Harloe, M., Issacharof, R. and Minns, R. (1974) *The Organisation of Housing*, Heinemann, London

Hartmann, C. (1964) 'The Housing of Relocated Families', *Journal of the American Institute of Planners*, *30*, 266–86

Hartman, G.W. and Hook, J.C. (1956) 'Substandard Urban Housing in the United States: A Quantitative Analysis', *Economic Geography*, *32*, 95–114

Harvey, D. (1973) *Social Justice and the City*, Arnold, London

Hawley, A.H. and Duncan, O.D. (1957) 'Social Area Analysis: A Critical Appraisal', *Land Economics*, *33*, 227–45

Headey, B. (1978) *Housing Policy in the Developed Economy*, Croom Helm, London

Helper, R. (1969) *Racial Policies and Practices of Real Estate Brokers*, Minneapolis University Press, Minneapolis

Herbert, D.T. (1967) 'Social Area Analysis: A British Study', *Urban Studies*, *4*, 41–66

Herbert, D.T. (1972) *Urban Geography: A Social Perspective*, David and Charles, Newton Abbot

Herbert, D.T. (1973) 'The Residential Mobility Process: some Empirical Observations', *Area*, *5*, 44–8

Hill, A.G. (1973) 'Segregation in Kuwait' in Clark, B.D. and Gleave, M.B. (eds.) *Social Patterns in Cities*, Institute of British Geographers Special Publication no. 3, London

Hole, W.V. (1959) 'Social Effects of Planned Rehousing', *Town Planning Review*, *30*, 161-73

Hole, W.V. (1979) 'Social Research in Housing: A Review of Progress in Britain since World War II', *Local Government Studies*, *5*, 23-40

Hoyt, H. (1939) *The Structure and Growth of Residential Neighbourhoods in American Cities*, Federal Housing Administration, Washington, DC

Jennings, J.H. (1971) 'Geographical Implications of the Municipal Housing Programme in England and Wales', *Urban Studies*, *8*, 121-38

Johnston, R.J. (1969) 'Some Tests of a Model of Intra-urban Population Mobility: Melbourne, Australia', *Urban Studies*, *6*, 34-57

Johnston, R.J. (1971) *Urban Residential Patterns: An Introductory Review*, Bell & Sons, London

Jones, E. (1961) *A Social Geography of Belfast*, Oxford University Press, Oxford

Kain, J.F. (1962) 'The Journey to Work as a Determinant of Residential Location', *Papers of the Regional Science Association*, *9*, 137-60

Kain, J.F. and Quigley, J.M. (1970) 'Measuring the Quality of the Residential Environment', *Environment and Planning*, *2*, 23-32

Kain, J.F. and Quigley, J.M. (1972) 'Housing Market Discrimination, Homeownership, and Savings Behaviour', *American Economic Review*, *62*, 263-77

Kasperson, R.E. (1971) 'The Post-behavioural Revolution Geography', *British Columbia Geographical Series*, *12*, 5-20

Kirby, A.M. (1976) 'Housing Market Studies: A Critical Review', *Transactions, Institute of British Geographers*, *1*, 2-7

Kirby, D.A. (1971) 'The Inter-war Council Dwelling: A Study of Residential Obsolescence and Decay', *Town Planning Review*, *42*, 250-68

Kirby, D.A. (1973) 'Residential Mobility among Local Authority Tenants', *Housing and Planning Review*, *29(2)*, 11-12

Kirby, D.A. (1974) 'Residential Growth: the Inter-war Years in England and Wales', *The Local Historian*, *11*, 24-30

Kirby, D.A. (1979) *Slum Housing and Residential Renewal: the Case in Urban Britain*, Longman, London

Kirwan, R.M. and Martin, D.B. (1972) *The Economics of Urban Residential Renewal and Improvement*, Centre for Environmental Studies, London

Kivell, P.T. (1975) 'Postwar Urban Residential Growth in North Staffordshire' in Philips, A.D.M. *Environment, Man and Economic Change: Essays Presented to S.H. Beaver*, Longman, London

Lambert, C. (1976) *Building Societies, Surveyors and the Older Areas of Birmingham*, Working paper 38, Centre for Urban and Regional Studies, University of Birmingham

Lambert, J., Paris, C. and Blackaby, R. (1978) *Housing Policy and the State*, Macmillan, London

Lansing, J.B., Clifton, C.W. and Morgan, J.N. (1969) *New Homes and Poor People: A Study of Chains and Moves*, Institute for Social Research, University of Michigan, Ann Arbor

Lansley, S. (1979) *Housing and Public Policy*, Croom Helm, London

Leslie, G.R. and Richardson, A.H. (1961) 'Life-cycle, Career Pattern and the Decision to Move', *American Sociological Review*, *25*, 894-902

Listokin, D. (1973) *The Dynamics of Housing Rehabilitation: Macro and Micro Analyses*, Centre for Urban Policy Research, Rutgers University

Lowry, I.S. (1964) *A Model of Metropolis, R.M.-4035-RC*, Rand Corporation,

Santa Monica
Lyon, S. and Wood, M.E. (1977) 'Choosing a House', *Environment and Planning*, 9, 1169–76
MacQueen, J. and Miller, R.G. Jr (1960) 'Optimal Persistence Policies', *Operations Research*, 10, 362–80
Maisel, S.J. (1966) 'Rates of Ownership, Mobility and Purchase' in *Essays in Urban Land Economics*, University of California Press, Berkeley
McElrath, D.C. (1962) 'The Social Areas of Rome: A Comparative Analysis', *American Sociological Review*, 27, 376–91
McGuire, C.C. (1981) *International Housing Policies: A Comparative Analysis*, Lexington Books
Mellor, R. (1973) 'Planning for Housing: Market Processes and Constraints', *Planning Outlook*, 13, 26–42
Michelson, W. (1977) *Environmental Choice, Human Behaviour and Residential Satisfaction*, Oxford University Press, New York
Montero, F.C. (1968) 'Social Aspects of Rehabilitation', *Building Research*, January–March, 17–21
Montgomery, D.S. (1960) 'Relocation and its Impact on Families', *Social Casework*, October, pp. 402–7
Morgan, B.S. (1976) 'The Bases of Family Status Segregation: A Case Study in Exeter', *Transactions, Institute of British Geographers*, 1, 83–107
Morris, P. (1962) 'The Social Implications of Urban Redevelopment', *Journal of the American Institute of Planners*, August, pp. 180–6
Morrison, P.A. (1967) 'Duration of Residence and Prospective Migration: the Evaluation of a Stochastic Model', *Demography*, 4, 553–61
Murdie, R.A. (1969) 'Factorial Ecology of Metropolitan Toronto', 1951–1961, *Research Paper 116*, Department of Geography, University of Chicago
Murie, A. (1975) *The Sale of Council Houses*, Occasional Paper 35, Centre for Urban and Regional Studies, University of Birmingham
Murie, A., Niner, P. and Watson, C. (1976) *Housing Policy and the Housing System*, Allen and Unwin, London
Muth, R.F. (1969) *Cities and Housing*, Chicago University Press
Needleman, L. (1965) *The Economics of Housing*, Staples Press, London
Needleman, L. (1968) 'Rebuilding or renovation – A Reply', *Urban Studies*, 5, 86–90
Niner, P. (1975) *Local Authority Housing Policy and Practice – A Case Study Approach*, Occasional Paper 31, Centre for Urban and Regional Studies, University of Birmingham
Openshaw, S. (1976) 'An Empirical Study of some Spatial Interaction Models', *Environment and Planning*, 8, 23–41
Pahl, R.E. (1970) *Whose City? And Further Essays on Urban Society*, Penguin Books, Harmondsworth
Palm, R. (1976) 'The Role of Real Estate Agents as Information Mediators in Two American Cities', *Geografiska Annaler*, 58B, 28–41
Palm, R. (1979) 'Financial and Real Estate Institutions in the Housing Market: A Study of Recent House Price Changes in the San Francisco Bay Area' in Herbert, D.T. and Johnston, R.J. (eds.) *Geography and the Urban Environment*, 2, John Wiley, Chichester
Paris, C. (1977) 'Housing Action Areas', *Roof*, 2, 9–13
Paris, C. and Lambert, J. (1979) 'Housing Problems and the State: the Case of Birmingham, England' in Herbert, D.T. and Johnston, R.J. (eds.) *Geography and the Urban Environment: Progress in Research and Applications 2*, John Wiley, Chichester
Park, R.E. (1952) *Human Communities*, New York

Park, R.E. (1961) 'Human Ecology' in Theodorson, G.A. (ed.) *Studies in Human Ecology*, Harper and Row, Evanston, Ill

Parker, J. and Dugmore, K. (1976) *Colour and the Allocation of GLC Housing: the Report of the GLC Lettings Survey, 1974-5*, Greater London Council, London

Pearson, P. and Henney, A. (1972) *Home Improvement, People or Profit?*, Shelter, London

Peet, R. (1977) *Radical Geography: Alternative Viewpoints on Contemporary Social Issues*, Methuen, London

Pines, D. (1975) 'On the Spatial Distribution of Households According to Income', *Economic Geography*, *51*, 142-9

Rainwater, L. (1966) 'Fear and the House-as-Haven in the Lower Class', *Journal of the American Institute of Planners*, *32*, 23-31

Rapoport, A. (1969) *House Form and Culture*, Foundations of Cultural Geography Series, Prentice-Hall, Englewood Cliffs

Rees, P.H. (1970) 'Concepts of Social Space' in Berry, B.J.L. and Hooton, F.E. (ed.) *Geographic Perspectives on Urban Systems*, Prentice Hall, Englewood Cliffs

Rex, J. (1968) 'The Sociology of a Zone in Transition' in Pahl, R.E. (ed.) *Readings in Urban Sociology*, Pergamon Press, Oxford

Rex, J. and Moore, R. (1967) *Race, Community and Conflict: A Study of Sparkbrook*, Oxford University Press, Oxford

Richardson, H.W. (1971) *Urban Economics*, Penguin, Harmondsworth

Roberts, J.T. (1976) *General Improvement Areas*, Saxon House, Farnborough

Robson, B.T. (1966) 'An Ecological Analysis of the Evaluation of Residential Areas in Sunderland', *Urban Studies*, *3*, 120-42

Robson, B.T. (1969) *Urban Analysis: A Study of City Structure with Special Reference to Sunderland*, Cambridge University Press, Cambridge

Robson, B.T. (1979) 'Housing, Empiricism and the State' in Herbert, D.T. and Smith, D.M. (eds.) *Social Problems and the City: Geographical Perspectives*, Oxford University Press, Oxford

Rossi, P.H. (1955) *Why Families Move*, The Free Press, New York

Schaaf, A.H. (1969) 'Economic Feasibility Analysis for Urban Renewal Housing Rehabilitation', *Journal of the American Institute of Planners*, *35*, 399-404

Schnore, L.F. (1963) 'The Socio-economic Status of Cities and Suburbs', *American Sociological Review*, *28*, 76-85

Schnore, L.F. (1964) 'Urban Structure and Surburban Selectivity', *Demography*, *1*, 164-76

Schnore, L.F. (1965) 'On the Spatial Structure of Cities in the Two Americas' in Hauser, P.M. and Schnore, L.F. (eds.) *The Study of Urbanisation*, John Wiley & Sons, New York

Schnore, L.F. (1966) 'Measuring City-Surburban Status Differences', *Urban Affairs Quarterly*, *1*, 95-108

Senior, M.L. (1973) 'Approaches to Residential Location Modelling 1: Urban, Ecological and Spatial Interaction Models (A Review), *Environment and Planning*, *5*, 165-97

Senior, M.L. (1974) 'Approaches to Residential Location Modelling 2: Urban Economic Models and some Developments (A Review)', *Environment and Planning*, *6*, 369-409

Shevky, E. and Williams, M. (1949) *The Social Areas of Los Angeles*, University of California Press, Los Angeles

Shevky, E. and Bell, W. (1955) *Social Area Analysis*, Stanford University Press, Stanford, California

Short, J.R. (1977) 'The Intra-urban Migration Process: Comments and Empirical

Findings', *Tijdschrif voor economische en sociale geographie*, *68*, 362–71

Short, J.R. (1978) 'Residential Mobility in the Private Housing Market of Bristol', *Transactions, Institute of British Geographers*, *3*, 533–47

Short, J.R. 'Landlords and the Private Rented Sector: A Case Study' in Boddy, M. (ed.) *Property, Investment and Land*, Working Paper 2, School for Advanced Urban Studies, Bristol

Short, J.R. and Bassett, K.A. (1978) 'Housing Action Areas: An Evaluation', *Area*, *10*, 153–7

Sigsworth, E.M. and Wilkinson, R.K. (1967) 'Rebuilding or Renovation', *Urban Studies*, *4*, 109–22

Simmons, J.W. (1968) 'Changing Residence in the City: A Review of Intraurban Mobility', *Geographical Review*, *58*, 622–51

Smailes, A.E. (1955) 'Some Reflections on the Geographical Description and Analysis of Townscapes', *Transactions, Institute of British Geographers*, *21*, 99–115

Smith, D.M. (1977) *Human Geography: A Welfare Approach*, Arnold, London

Speare, A., Goldstein, S. and Frey, W.H. (1975) *Residential Mobility, Migration and Metropolitan Change*, Ballinger, Cambridge, Mass.

Spencer, K.M. (1970) 'Older Urban Areas and Housing Improvement Policies', *Town Planning Review*, *41*, 250–62

Stafford, D.C. (1978) *The Economics of Housing Policy*, Croom Helm, London

Stone, M.E. (1978) 'Housing, Mortgage Lending, and the Contradictions of Capitalism', in Tabb, W.K. and Sawers, L. (eds.) *Marxism and the Metropolis*, Oxford University Press, New York

Taylor, P.J. (1978) *'Difficult to Let, Difficult to Live in and sometimes, Difficult to get out of': An Essay on the Provision of Council Housing with Special Reference to Killingworth'*, Discussion Paper 16, Centre for Urban and Regional Development Studies. University of Newcastle-upon-Tyne

Thomas, C.J. (1966) 'Some Geographical aspects of Council Housing in Nottingham', *The East Midland Geographer*, *4*, 88–98

Timms, D. (1971) *The Urban Mosaic: Towards a Theory of Residential Differentiation*, Cambridge University Press, Cambridge

Troy, P.N. (1973) 'Residents and their Preferences: Property Prices and Residential Quality', *Regional Studies*, *7*, 183–92

Van Ardsol, Jr M.D., Camilleri, S.F. and Schmid, C.F. (1958) 'The Generality of Urban Social Area Indexes', *American Sociological Review*, *23*, 277–84

Vitarello, J. (1975) 'The Redlining Route to Urban Decay', *Focus*, *3*, 4–5

Wagner, P.L. (1969) 'Editorial Introduction to Rapoport, A., *House Form and Culture*' Foundations of Cultural Geography Series, Prentice Hall, Englewood Cliffs

Watson, C.J. (1973) *Household Movement in West Central Scotland: A Study of Housing Chains and Filtering*, Occasional Paper, no. 26, Centre for Urban and Regional Studies, University of Birmingham

Weir, S. (1976) 'Redline Districts', *Roof*, *1*, 109–14

Wendt, P.F. (1963) *Housing Policy – the Search for Solutions: A Comparison of the United Kingdom, Sweden, West Germany and the United States since World War II*, University of California Press, Berkeley

Whitehand, J.W.R. (1967) 'The Settlement Morphology of London's Cocktail Belt', *Tijdschrift voor Economische en Sociale Geografie*, *58*, 20–7

Whitehead, C.M.E. (1975) *A Model of the New Housing Market in Great Britain*, Saxon House, Farnborough

Whyte, W. (1960) *The Organisation Man*, Penguin Books, Harmondsworth, Middlesex

Wilkinson, R. and Merry, D.M. (1965) 'A Statistical Analysis of Attitudes to

Moving', *Urban Studies*, *2*, 1–14

Williams, P. (1976a) *The Role of Financial Institutions and Estate Agents in the Private Housing Market: A General Introduction*, Working Paper 39, Centre for Urban and Regional Studies, University of Birmingham, Birmingham

Williams, P. (1976b) 'The Role of Institutions in the Inner London Housing Market: the Case of Islington', *Transactions, Institute of British Geographers*, *1*, 72–82

Williams, P. (1978) 'Building Societies and the Inner City', *Transactions, Institute of British Geographers*, *3*, 23–34

Wilson, A.G. (1974) *Urban and Regional Models in Geography and Planning*, Wiley, London

Wilson, R. (1963) *Difficult Housing Estates*, Pamphlet no. 5, Tavistock Institute

Wolman, H.L. (1975) *Housing and Housing Policy in the US and the UK*, Lexington Books, Lexington, Mass.

2 EMPLOYMENT AND UNEMPLOYMENT

P.J. Bull

From the extensive media coverage of the current economic and social problems of large cities in the UK, and especially of the so-called inner-city problem, one may have received the distinct impression that their high levels of unemployment and large employment losses began in the 1970s. Such a conclusion would be a mistake. Admittedly the decline of employment opportunities in the conurbations is of a recent nature, beginning in the early 1960s (Drewett *et al.*, 1975), but the high levels of unemployment, and their closely associated problems of low incomes and bad housing, have been a constant feature of the British urban scene since at least the beginning of the nineteenth century (Engels, 1969; Stedman Jones, 1971). This discussion of employment and unemployment in cities, however, will concentrate on the post-war period, and in particular on the last 25 years. Most of the evidence will be drawn from research in the UK but important material from other western nations will also be cited. The chapter has two principal aims. First to describe the changing inter- and intra-urban patterns of employment and unemployment, and second to begin to erect a possible explanation for these changes. The chapter will conclude with a discussion of some policy issues. Finally, by way of introduction it must be noted that although services now employ more people than manufacturing in all the conurbations of the UK, this chapter will place most emphasis on manufacturing industry. There are four reasons for this emphasis. First there is a wide and varied literature to consider on urban manufacturing activity. This is not the case for the service sector. Second, recent changes in urban manufacturing activity have been far more dramatic than in services. Third, spatial changes in services such as transport, distribution and retailing are relatively easy to account for because they are so closely linked to population movements (Keeble, 1978). Finally, many of the recent policy initiatives designed to promote the economy of the central areas of cities have tended to concentrate on the manufacturing sector.

Inter-urban Changes in Employment and Unemployment

Since the end of the 1950s there has been a marked change in the geography of employment in the UK. The conurbations have lost large numbers of jobs while the smaller free standing towns and cities and many of the settlements in the more rural areas of the country have recorded employment gains.

In Table 2.1, derived from the work of Fothergill and Gudgin (1979a), employment change between 1959 and 1975 in the 62 sub-regions of the UK has been divided into eight categories along an urban-rural continuum. From this table it can be seen that the decline of London and the major conurbations by over 750,000 jobs was particularly dramatic. However, the decline in the manufacturing sector was even more severe, with London losing nearly 40 per cent of its 1959 total, and with the other conurbations contracting by over 400,000 jobs. In addition, it is clear from the work of Keeble (1976; 1980) that the period of greatest employment contraction in these areas was during the latter part of this 15 year period. For example, between 1959 and 1966 the West Midlands and Merseyside conurbations recorded modest increases in their manufacturing employment totals, yet by 1975 the former had lost over 100,000 jobs.

It can also be seen from Table 2.1 that large numbers of new jobs were generated in the major free-standing cities such as Southampton and Bristol, in the smaller free-standing cities including Oxford and Norwich, and in the towns of the industrial non-city group of sub-regions which consist principally of the coalfield areas of the North West, Wales and Yorkshire. In the case of the manufacturing sector it is only settlements in this latter group which recorded any major absolute job increases. On the other hand, in percentage terms, only settlements in the more rural parts of Britain, where few manufacturing jobs existed in 1959, recorded large growth rates. However, one must not conclude from this result that all rural settlements were growing during the 1960s and early 1970s. Even in one of the most successful employment generating regions of this period, East Anglia, there was great diversity in the growth rates of individual settlements. Between 1961 and 1971 the larger towns in this area such as Ipswich, Cambridge and Great Yarmouth were indeed expanding, whereas many of the smaller and more inaccessible settlements were in decline (Moseley and Sant, 1977). Nevertheless, although since the first quarter of the 1960s there has been a strong spatial shift in the location of employment away from the conurbations, these relatively small spatial units

Table 2.1: Employment Change by Type of Sub-region in the UK, 1959-75

(a) Total Employment

	1959 '000s	1975 '000s	Absolute change '000s	Percentage change	% UK employment 1959	% UK employment 1975
London	4519.3	4002.2	−517.1	−11.4	21.2	18.0
Conurbations	5524.6	5266.4	−258.2	−4.7	26.0	23.6
Major free standing cities	4426.3	4864.0	437.7	9.9	20.8	21.8
Smaller free standing cities	1230.8	1440.5	209.7	17.0	5.8	6.5
Industrial non-city	3712.4	4529.4	816.7	22.0	17.4	20.3
Urban non-industrial	510.9	607.2	96.3	18.8	2.4	2.7
Semi-rural	694.4	826.1	131.7	19.0	3.3	3.7
Rural	658.7	753.2	94.5	14.3	3.1	3.4
Total	21277.4	22289.0	1011.6	4.8	100.0	100.0

(b) Manufacturing Employment

	1959 '000s	1975 '000s	Absolute change '000s	Percentage change	% UK employment 1959	% UK employment 1975
London	1551.4	965.5	−585.9	−37.8	18.4	12.1
Conurbations	2733.7	2300.0	−433.7	−15.9	32.5	28.8
Major free standing cities	1833.9	1895.5	61.6	3.4	21.8	23.8
Smaller free standing cities	357.9	421.9	64.0	17.9	4.3	5.3
Industrial non-city	1567.7	1822.3	254.8	16.3	18.6	22.9
Urban non-industrial	107.6	149.3	41.7	38.8	1.3	1.9
Semi-rural	174.0	252.2	78.2	44.9	2.1	3.2
Rural	94.7	167.8	73.1	77.2	1.1	2.1
Total	8420.9	7974.5	−446.4	−5.3	100.0	100.0

(c) Service Employment[a]

	1959 '000s	1971 '000s	Absolute change '000s	Percentage change	% UK employment 1959	% UK employment 1971
London	2588.2	2647.1	58.9	2.3	27.0	24.9
Conurbations	2144.7	2256.8	112.1	5.2	22.4	21.2
Major free standing cities	1798.7	2088.3	289.6	16.1	18.8	19.6
Smaller free standing cities	619.7	708.6	88.9	14.3	6.5	6.7
Industrial non-city	1399.0	1800.1	401.1	28.7	14.6	16.9
Urban non-industrial	313.8	354.9	41.1	13.1	3.3	3.3
Semi-rural	350.7	399.1	48.4	13.8	3.7	3.8
Rural	372.0	376.6	4.6	1.2	3.9	3.5
Total	9586.8	10631.5	1044.7	10.9	100.0	100.0

Note: a. 1959–71.
Source: Derived from Fothergill and Gudgin (1979a) Tables 3.1, 3.2, 3.3, 3.7 and App. 2.

still dominated the geography of employment in the UK in the mid-1970s by possessing over 40 per cent of all jobs.

The information on employment in services by type of subregion in Table 2.1, unfortunately only for the shorter period 1959 to 1971, displays similar trends to those already outlined for all employment and manufacturing employment. The dominance of London and the conurbations is clearly apparent with the capital accounting for approximately 25 per cent of all employment in this sector throughout the period. There is also in the share of service employment by subregion type, a definite trend away from these settlements to the larger and smaller free-standing cities, and to the industrial non-city subregions. However, by 1971 this growth impulse had not moved sufficiently down the urban-rural continuum of subregions to inflate significantly either the growth rates or the shares of service employment in the non-urbanised and rural parts of the country. Nevertheless, Lloyd and Reeve (1981) have shown, with evidence for the later period 1971 to 1977, that for at least one of the planning regions of the UK, namely the North West, this shift or growth in service employment to smaller settlements continued throughout most of the 1970s. Indeed they also discovered that service employment in one of the two conurbations in their study area, Merseyside, began to decline.

Absolute and relative shifts in the distribution of employment away from the major cities is, of course, not peculiar to the UK. It characterised many western countries during the last twenty years. For example, in the USA between March 1962 and March 1978 settlements in areas outside the 225 Standard Metropolitan Statistical Areas (SMSAs) accounted for 56 per cent (1.8 million jobs) of the nation's increase in manufacturing employment. Furthermore, by 1970 the SMSAs had started to decline and over the next eight years lost approximately 500,000 manufacturing jobs, while the towns in the non-metropolitan areas gained almost 600,000 (Haren and Holling, 1979).

From the above discussion it may be tempting to suggest that the net result of the shift in the location of employment expansion away from the major urban areas to the smaller free-standing cities and towns should be an increase in the unemployment rates in the former and a decrease in the latter. Unfortunately, the evidence that is available for the UK for the 1963–75 period (Frost and Spence, 1981) and between February 1980 and February 1981 (Gillespie and Owen, 1981) tends to suggest no relationship between the magnitude of male unemployment or its change through time and an urban-rural continuum. There are probably two main reasons for this seemingly

Table 2.2: Male Unemployment Rates in the British Conurbations in 1961, 1971 and 1981

| | % unemployment rate | | | % of GB rate | | |
	1961[a]	1971[a]	1981[b]	1961	1971	1981
Greater London	2.4	4.7	7.4	80	87	64
West Midlands	2.4	5.2	13.0	80	96	112
South East Lancashire	3.1	6.2	11.7	103	115	101
West Yorkshire	2.6	5.8	11.3	87	107	97
Tyneside	5.0	9.7	14.3	167	180	123
Merseyside	5.8	9.4	17.5	193	174	151
Central Clydeside	6.1	11.0	15.5	203	204	134
Great Britain	3.0	5.4	11.6	100	100	100

Source: a. Corkindale (1976) annex, table 15. Residential unemployment rates recorded by the Census of Population. They refer to the number of registered unemployed or out of employment in an area as a percentage of the economically active residents.
b. Gillespie and Owen (1981). Rate of registered unemployed, February 1981.

surprising result. First, population movements, including planned migration to New and Expanded towns, have tended to go hand-in-hand with the spatial redistribution of employment opportunities. As a result the availability of new jobs in some towns may have been matched by in-migration. Second, the creation of many new jobs for women in both manufacturing and services has brought many people into employment who previously may not have been on the unemployment register. Thus new jobs may be filled with no impact on the number registered as unemployed.

Male unemployment rates for the seven British conurbations in 1961, 1971 and 1981 are presented in Table 2.2. Although one must be cautious in their interpretation because they stem from different sources, these results show a marked increase in the rate of male unemployment in all the conurbations over the 20 year period. However, they also show that with the significant exception of the West Midlands most conurbations improved their unemployment positions *relative* to national trends. This was most apparent for Tyneside, Merseyside and Clydeside where the highest rates of male unemployment were recorded throughout the 20 year period. Thus, it would appear that one of the effects of the recent severe recession has been a convergence of conurbation unemployment rates, with male unemployment rates increasing at a faster rate in the conurbations with relatively low rates in 1961. However, in general, one must conclude that, with the one notable exception of London, male unemployment rates for the largest UK cities have tended to be higher than for GB as a whole. Nevertheless, there is no evidence to suggest that the rate of unemployment is directly related to an urban-rural continuum.

Intra-urban Employment Change

The principal recent spatial change in employment opportunities in most cities in western societies has been one of decentralisation, or suburbanisation. That is a net shift of industrial activity away from the central areas of cities. Evidence on the suburbanisation of employment in the Metropolitan Economic Labour Areas (MELAs) of GB for the 1950s and 1960s can be found in Drewett *et al.* (1976). Their findings, summarised in Table 2.3, show that both the urban cores and metropolitan rings expanded during the 1950s and that a small increase in the spatial concentration of employment in the former areas took place. However, for the following decade the situation was reversed. An absolute decentralisation trend is apparent with the core areas recording an employment decline of over 3.0 per cent while the metropolitan rings expanded by 15.0 per cent. It can also be seen from Table 2.3 that, even after this degree of spatial change during the 1960s, the urban cores with almost 60 per cent of total MELA employment in 1971 still dominated the geography of employment in British cities. Similar decentralisation trends are also apparent from research in the USA. For example, for the 40 largest SMSAs between 1948 and 1963 Kain (1968) has demonstrated the existence of substantially higher rates of employment increase in suburban rather than in central areas. From Table 2.4 one can see this trend clearly for the four industrial sectors investigated. The two most important employing sectors, manufacturing and retailing, recorded absolute employment losses in the central cities in two of the three 5-year time periods, while all sectors, and in particular wholesaling and services, expanded in the suburbs throughout the period. Furthermore in contast to GB, the post 1950 suburbanising trend of manufacturing employment in the USA had been so profound that by 1967 the suburban rings of the 245 SMSAs accounted for a larger share of metropolitan manufacturing employment than their central cities (Dicken and Lloyd, 1981).

Of the voluminous available literature on the locational dynamics of industrial activity in urban areas, the majority of the recently published research has been primarily concerned with the manufacturing sector in the major urban areas. Some surveys of the changing structures and

Table 2.3: Employment Change in the Metropolitan Economic Labour Areas of GB, 1951–71

| | 1951–61 | | 1961–71 | | 1971 | |
	absolute (000s)	%	absolute (000s)	%	absolute (000s)	%
Urban cores	902	6.7	−439	−3.1	13898	58.6
Metropolitan rings	293	6.6	707	15.0	5429	22.9
Outer Metropolitan rings	−14	−0.4	130	3.9	3443	14.5
Unclassified areas	−56	−5.5	−7	−0.7	963	4.1
						100.0

Employment change

Source: Drewett *et al.* (1976), p. 16.

patterns of manufacturing in smaller cities are available including Roanoke, Virginia (Stuart, 1968) and Columbus, Ohio (Cowan, 1971) in the USA and Stoke-on-Trent (Whitelegg, 1976) and Northampton (Roberts, 1979) in the UK. However, very little detailed research has been produced on the tertiary sectors of cities, although notable exceptions include studies of office construction in New York between 1960 and 1975 (Schwartz, 1979), the decentralisation of office employment in Toronto, 1956-70 (Gad, 1979) and the development of the office sector in Manchester in the late 1960s and early 1970s (Damesick, 1979).

Much of this recent work on manufacturing in large urban areas has resulted directly from the setting up of establishment based data banks. These data sets, often including information on many thousands of manufacturing plants and requiring the input of many hundreds of man hours for their completion (Swales, 1976), have permitted a detailed analysis of the many individual locational events generating changes in the geography of manufacturing activity in cities. In the investigations using such establishment scale data, spatial change is regarded as the result of a number of *components of change*.

(1)   Plant births or openings resulting from
    (i)   the birth of completely new plants
    (ii) the transfer of a plant into the study area.
(2)   Plant deaths or closures, resulting from
    (i)   the complete closure, or death, of plants
    (ii) the transfer of a plant out of the study area.
(3)   Intra-urban transfers.

Table 2.4: Mean Annual Percentage Change in Employment in the 40 Largest SMSAs in the USA, 1948-63

	Central City[a]				Suburban Ring			
	Mean employment in 1948	Mean annual percentage change			Mean employment in 1948	Mean annual percentage change		
		1948-54	1954-8	1958-63		1948-54	1954-8	1958-63
Manufacturing	118,652	1.9	−1.7	−0.4	58,805	13.2	6.9	6.0
Wholesaling	33,124	0.8	0.2	−0.2	2,959	24.9	16.6	15.1
Retailing	61,048	−0.6	0.1	−2.0	19,992	11.3	13.5	13.4
Services	23,654	1.6	3.9	0.9	4,240	18.0	16.6	13.5

Note: a. 1950 central city definition.
Source: Kain (1968) pp. 12 and 13.

(4)   Plant survivors, or plants remaining in the same location during the study period.

It is also important to note that in the last two categories changes in the other characteristics of manufacturing plants, such as number employed and net output, may also take place.

The cities for which investigations adopting this kind of establishment accounting procedure have been published include London (Dennis, 1978; Gripaios, 1977a and 1977b), Greater Manchester (Lloyd and Mason, 1978; Mason, 1980a and 1981), Merseyside (Lloyd, 1979), Clydeside (Cameron, 1973; Firn and Hughes, 1974; Bull, 1978), the West Midlands (Beasley, 1955; Johnson, 1958; Firn and Swales, 1978) and Leicester (Fagg, 1973; Gudgin, 1978) in the UK, and New York (Leone, 1971; Kemper, 1973), New Orleans (Saussy, 1972) and Cleveland, Minneapolis-St Paul, Boston and Phoenix (Struyk and James, 1975) in the USA.

One of the most important findings of this research has been the discovery of very high rates of locational change in all cities. For example in Clydeside between 1958 and 1968 a net loss of 182 establishments took place from a base of 2,505 in 1958. This loss reflected gross changes of 711 plant deaths and 529 plant births. In addition, 607 plants changed their location in the conurbation during the 11 years (Bull, 1978). Thus 52.6 per cent of the plants in existence in 1958 had undertaken some form of locational change by the end of 1968. The equivalent rate for Merseyside between 1966 and 1975 was 59.8 per cent (Lloyd, 1979). In cities in the USA high component rates of change have also been discovered. For example, in the New York SMSA within a period of 25 months between 1967 and 1969 a plant death rate of 7.6 per cent, a birth rate of 10.2 per cent, and an intra-urban transfer rate of 11.5 were recorded from a base of 39,128 establishments in 1967 (Leone, 1971). Similarly in the New Orleans SMSA, for the period 1965 to 1969, the death, birth and intra-urban transfer rates were 21.5 per cent, 21.0 per cent and 15.1 per cent respectively (Saussy, 1972).

As a result of these high establishment turnover rates locational change in employment may also be generated. Struyk and James (1975) provide evidence from which employment component rates of change may be calculated for four very different cities in the USA, namely Cleveland, Minneapolis-St Paul, Boston and Phoenix. From Table 2.5 in which these rates are presented, a number of interesting conclusions may be drawn. First, that in comparison to the plant

Table 2.5: The Composition of Manufacturing Employment Change, 1965–8, in Cleveland, Minneapolis-St Paul, Boston and Phoenix

| | Manufacturing employment 1965 | Percentage change 1965–8 | Employment Component Rates of Change (%)[a] | | | |
			Intra-urban transfers[b]	Births	Deaths	Net change in survivors
Cleveland	297,992	-4.7	5.8	2.6	7.8	0.4
Minneapolis-St Paul	151,591	-5.7	8.2	6.1	11.2	0.9
Boston	298,118	3.8	4.7	1.3	8.0	11.0
Phoenix	52,816	27.1	4.7	12.1	5.3	20.4

Notes: a. Percentage of 1965 employment base.
b. Using employment at transfer origin.
Source: Struyk and James (1975) derived from Tables 2.7, 3.6, 4.6 and 5.5.

component rates for New York and New Orleans given above these results appear to be much lower. Clearly such results are dependent upon the mix of plant sizes within individual cities but, as we shall see below, it also points to the large numbers of small plants involved in locational changes in most large urban areas. Second, the results for Boston demonstrate that it does not always follow that because a city is expanding in employment terms its employment birth rate will be high and exceed its death rate. Indeed, the employment birth rates in the two employment shedding economies exceeded the low birth rate of only 1.3 per cent for Boston. Third, and related to this last point, is the profound impact that *in-situ* employment change can have in generating growth of manufacturing activity in an urban economy. Phoenix, for example, recorded an employment increase of 20 per cent due to net *in-situ* change in surviving plants. This is in stark contrast to the net changes of less than one per cent in the two labour shedding urban areas. It is also important to realise that these figures relate to *net* change. Gross employment changes may be substantially greater. In Greater Manchester, for example, between 1966 and 1975 net employment change in surviving establishments accounted for a 4.3 per cent decline of 19,433 jobs. However, this was the result of a loss of 64,720 jobs and an addition of 45,287 (Mason, 1980a).

Mason (1980a) has also provided evidence to indicate marked variations in the opening and closure employment rates between some of the conurbations in the UK. For example, for establishments with more than 5 employees, Greater Manchester recorded an opening rate of 5.8 per cent and a closure rate of 27.1 per cent between 1966 and 1971, while Clydeside recorded equivalent rates of 9.8 per cent and 16.4 per cent between 1958 and 1968. Similarly for all plants with 20 or more employees the opening and closure rates for Greater London were 1.0 per cent and 21.7 per cent between 1966 and 1974 and for Merseyside 19.8 per cent and 16.9 per cent between 1966 and 1975. These variations conform to a Development Area/non-Development Area division. The two conurbations with relatively high opening rates and low closure rates, Clydeside and Merseyside, have enjoyed Development Area (DA) status for at least the last 30 years, whereas Greater Manchester only began to receive regional aid in 1972 as part of an Intermediate area, and Greater London has never benefited from such schemes. Indeed Greater London has until recently been regarded as one of the areas from which employment can be diverted to the DAs. Thus, the operation of regional policy may in part help to account for some of the differences in employment component rates between conurbations in the UK. While conurbations such as Clydeside and Merseyside have clearly gained employment from the operation of regional policy, principally through the creation of branches by companies based in the South East and Midlands, London and the West Midlands conurbation have lost employment. However, one must not conclude from these findings that the operation of regional policy has been the principal reason for the decline of manufacturing employment in either of these two areas. For example, Dennis (1978) has shown for the case of Greater London between 1966 and 1974 that industrial movement from this area, either through the closure of London plants and the transfer of production to locations elsewhere, or through employment shifts from surviving London plants to other branches, accounted for not more than 105,300 manufacturing jobs, or 27 per cent of the total decline of 390,000. Furthermore, only 36,200 or 9 per cent, was due to movement to the DAs. The most important component of the decline of manufacturing employment in London during this time period, which accounted for 44 per cent of the total employment loss, was the net decline due to the difference between openings and complete closures unassociated with movement.

Suburbanisation

The components of change approach has been particularly useful in advancing our understanding of the ways in which the suburbanisation or decentralisation of manufacturing activity has taken place. Suburbanisation can be promoted in three ways:

(1)   The transference of plants from central to peripheral parts of cities.
(2)   A higher plant closure rate in central locations than in the suburbs, and/or a higher opening rate in the suburbs.
(3)   Surviving or stationary plants in the suburbs growing faster, or declining more slowly, than those in central areas.

In all cities decentralisation has been promoted to a lesser or greater degree by each of these mechanisms, and for some urban areas it has been possible to isolate the most important ones.

The one consistent finding from all the investigations of the intra-urban spatial changes of manufacturing activity over the last 20 years has been the minor role played by transfers in the decentralisation process. For example, of the 33,021 jobs lost from the central 25 kilometre squares of Greater Manchester between 1966 and 1975 only 5,209 or 15.8 per cent, were due to the intra-urban transference of establishments (Mason, 1981). Similarly in Clydeside between 1958 and 1968 only 36 per cent of the 407 plant transfers originating within 4 kilometres of a predefined conurbation centre point recorded destinations outside this area (Cameron, 1980). Further evidence supporting this view is available for New York (Leone, 1971), Greater Leicester (Fagg, 1973), Merseyside (Lloyd, 1979) and Greater London (Dennis, 1978). Thus, although it is possible that during earlier time periods transfers may have played a significant role in the decentralisation of manufacturing activity, this has certainly not been the case in the post-1960 era (Wood, 1974).

For most UK cities the most potent reason for suburbanisation has been the spatial variation in establishment opening and closure rates with distance from urban cores. In most cities establishment and employment opening rates have tended to increase with distance from the city centre. For example, in Clydeside between 1958 and 1968 establishment birth rates in the inner area, outer central area and non-urban zone were 14.8 per cent, 21.3 per cent and 41.1 per cent respectively (Cameron, 1980). The one major exception to this finding was in Greater Manchester where employment opening rates were found to be relatively invariant with respect to distance from the conurbation centre. In this case suburbanisation was more the result of a marked gradation of employment closure rates: from 40 per cent within two kilometres of the central point to 25 per cent in the non-urban areas (Mason, 1981). Curiously in Greater London (Dennis, 1978) and Clydeside (Bull, 1978) employment and establishment closure rates respectively were found to be spatially invariant. Thus it would appear that for the UK cities investigated so far only Leicester (Fagg, 1973 and 1980) and Merseyside (Lloyd, 1965, 1970 and 1979) have shown signs of both the spatial variations in birth rates and death rates promoting decentralisation. Interestingly in this last conurbation *in-situ* employment change in surviving plants played a far larger part in the decline of the central area than in any other UK city: out of a net loss of 18,291 jobs between 1966 and 1975 net *in-situ* shrinkage accounted for 9,167 (Lloyd, 1979).

A number of studies have actually measured suburbanisation in terms of an increase in the average distance of plants and employment from a pre-defined and fixed central city location (Tulpule, 1969;

Bull, 1978; Mason, 1981). Perhaps one of the most surprising results to come from such analyses is the very small distances over which manufacturing activity as a whole has moved away from central city locations, despite high component rates of change. For example, in Clydeside (for which the establishment component rates have already been given), the median distance of plants from the conurbation centre increased by less than one kilometre between 1958 and 1968 (Bull, 1978). The principal reason for a result such as this is that much of the locational change recorded in all cities takes place within the existing stock of industrial premises. Thus, the spatial patterns of manufacturing industry in a city may remain relatively stable over a period of say 10 or 15 years although the individual establishments and enterprises within them may be changing very rapidly indeed. Furthermore, it signals the fact, about which more will be made in the next section, that some types of manufacturing plants and industries have failed to promote suburbanisation in the post-war period by continuing to display a distinct locational preference for central urban sites.

Some Possible Explanations

Although many descriptive accounts of the intra- and inter-urban changes of manufacturing activity have been published very few attempts have been made to explain these changes. The attempts that have been made tend to fall into one of two categories. First those based on the industrial structures of different areas, and second those concerned with the changing characteristics of plants as operating costs are minimised. Since it is extremely difficult to obtain detailed information on either intra-urban cost surfaces or the operating costs of individual establishments and firms the paucity of explanations of the second kind can be readily understood.

Industrial Structure. One possible approach to an explanation of the decline of both large cities and their inner areas might be to suggest that their principal problem lies in an initial possession of high levels of employment in industries that subsequently went into decline. If an explanation of this nature based on the industrial structures of areas was found to have some validity then it would tend to suggest that there were few explicitly locational reasons for the economic failure of these areas during the last 20 or 30 years. It is of course true that many large cities have suffered savage employment losses as individual industries have declined. For example, in the UK Tyneside and Clydeside have been adversely affected by the decline of British shipbuilding, Greater Manchester and West Yorkshire by the problems of the textile and clothing industries and, more recently, the West Midlands by the rationalisation of the British motor vehicle industry. Clearly a city will be adversely affected by a fall in the demand for the products in which its businesses specialise. However, most cities by virtue of their size tend to be highly diversified economies (Norcliffe, 1975). The diversified nature of the conurbations in the UK in contrast to the rest of the country has been demonstrated by Chisholm and Oeppen (1973). In addition, the inner areas of these conurbations are also characterised by high levels of industrial diversification. For example, in an area of just 30 square kilometres in central Manchester all the 17 manufacturing orders of the 1968 Standard Industrial Classification were represented in both 1966 and 1975 except shipbuilding and marine engineering (Dicken and Lloyd, 1979). Thus, it is most unlikely that the decline of any one or two industries could account for the decline of any city or its inner area. Furthermore, given the diversity of these areas one would expect them to be able to benefit from expanding industries. However, as Danson *et al.* (1980) have shown the employment growth of the central areas of the UK conurbations has fallen far short of that expected from their favourable industrial structures. From a shift-share analysis of all employment divided into 116 sectors they demonstrated that, given national growth rates and their individual employment structures, the inner areas between 1952 and 1976 declined by over 2.5 million jobs more than expected. Over the same time period the outer metropolitan areas grew by almost half a million jobs more than expected. Fothergill and Gudgin (1979a) have also shown for the period 1959–71 that industrial structure does not explain the poor performance of the conurbations and the success of the free-standing cities and towns. It is therefore clear that in the UK at least industrial structure has little to offer in helping to account for the decline of industrial activity in either the major cities or their inner areas.

The Changing Characteristics of Manufacturing Industry. In attempts to remain profitable the characteristics of manufacturing activity in western societies have changed substantially in the post-war period. These changes may be considered in terms of the adoption and exploitation of new technologies (Goddard and Thwaites, 1980) of which two are particularly relevant to this discussion, namely methods of production and communication.

(1)   Changes in production techniques to:

(a) more space-extensive production line systems requiring single-storey factories and storage facilities.

(b) larger factories as technological change permits the exploitation of scale economies.

(c) the substitution of capital for labour, and where the technology permits, the replacement of men by women as labour costs rise as a proportion of net output.

(2)   Changes in techniques of communication

(a) Improvements in road transport particularly through the building of motorways has helped to lead to a reduction in transportation costs as a proportion of net output and a greater flexibility in delivery locations and load sizes. International improvements in the movement of goods through for example containerisation have also led to significant savings in journey costs and times. These developments have helped to permit an increase in market areas, the specialisation of plants, and the interplant movement of components and sub-assemblies.

(b) Improvements in inter-personal contact systems — telephone, telex, etc. — permitting the strategic control of a number of different manufacturing units from one central headquarters.

All the above changes have helped to permit the growth of multi-unit firms and multi-firm corporations (Prais, 1976). Firms can now benefit from growth economies (Penrose, 1963) by setting up branches and by acquiring new firms and plants and still exercise a high degree of control over their activities. As a result, it is possible for traditional urban external economies to be replaced by in-house linkages between the existing units of firms for raw materials, components, information and many service functions (Lever, 1974). Furthermore, the success of any plant in a firm may no longer be measured in terms of its individual profitability but by the way it fits into the overall plans of a company or corporation.

The possible locational ramifications of these changes could include:

(1)   An increased demand for larger sites and premises. Thus the places where single-storey modern factory space is available and where land is relatively cheap become particularly important.

(2)   A reduced need for the spatial proximity of establishments. If the deliveries between plants are large, regular and planned then spatial proximity reduces in significance. It is only when

the contacts between businesses are at irregular intervals, necessitating speedy reactions to demand changes that close association becomes vital (Hoare, 1975). In addition, if production is on a relatively large scale and markets extensive then the inter-urban transportation of goods will become increasingly important. Therefore proximity to motorway access points may become significant.

As a result of both these points it is clear that the central areas of cities, often characterised by congested roads, high land values (Vallis, 1972; Royal Town Planning Institute, 1978) and technologically obsolete industrial premises, may have become increasingly unattractive locations for manufacturing activity in the post-war period. The required cheap land for space extensive production techniques can be found in the suburbs of most cities and in non-metropolitan areas. In the UK the construction of New Towns and suburban industrial estates in the cities benefiting from regional aid has helped to provide suitable premises. In addition, given the improvements in inter-urban communication networks modern manufacturing now possesses a high degree of locational flexibility. Thus locations in non-metropolitan areas where labour may be relatively cheap (Hoch, 1972) may be considered, and when many locations present similar cost structures non-pecuniary factors such as amenity and residential preferences (Keeble, 1976) may become important.

Thus it would seem reasonable to argue that as manufacturing activity has become increasingly orientated towards highly capitalised branch plant production in space extensive factories with fewer local (short distance) linkages and perhaps a greater proportion of female labour it would tend *not* to locate in:

(1)   the major metropolitan areas, and
(2)   the central areas of cities.

In terms of the spatial changes of manufacturing activity in cities, during periods of economic growth new factory space will be demanded with the characteristics described above leading to suburbanisation by the birth of plants in non-central locations, especially when the opportunities for *in-situ* expansion are limited. On the other hand, during periods of recession it may be the most modern factories which stand the greatest chance of survival, and as a consequence suburbanisation may take place through high inner-area death rates and labour shedding in the larger central establishments.

Due principally to data limitations the above ideas have not been comprehensively tested. However, for both the intra- and inter-urban cases there is some supporting evidence. In terms of the decline of the major cities of the UK Massey and Meegan (1978) have shown that for profitability to be maintained in 25 large firms in the electrical and electronic engineering industry between 1966 and 1972 large capacity cutbacks were necessary. This rationalisation resulted in the closure and contraction of the most labour intensive plants located in the cities of London, Liverpool, Manchester and Birmingham. Furthermore, only 10 per cent (433 jobs) of the employment transferred to new locations remained in these cities, the balance going to smaller settelements particularly in the Development Areas. The relative unprofitability of manufacturing activity in large urban areas has also been strongly suggested by the work of Lever (1982). By comparing the operating costs of a sample of firms in Clydeside in 1979 with national equivalents from the Census of Production operating costs were shown to be significantly higher in Clydeside than in the nation as a whole. The principal reason for this difference lay in high labour costs and inadequate capital investment. Thus, it may be very tentatively argued that investment outside the conurbations may be offsetting labour costs by higher levels of capital expenditure.

In the case of the suburbanisation of manufacturing activity there are numerous pieces of corrobative evidence for some of the propositions. For example, the tendency for branch plants to locate in the suburbs is a well known phenomenon, and the link between industrial growth and suburbanisation has been established with evidence from Clydeside (Firn and Hughes, 1974; Cameron, 1980). However, no formal testing of the whole structure has been undertaken and therefore no matter how plausible the ideas may appear they must be regarded at this stage solely as hypotheses.

The above model of the suburbanisation of manufacturing activity is designed to account for the most important secular trend in the geography of manufacturing activity in urban areas in a general way. Clearly it has little to offer in trying to account for the high establishment component rates of change recorded for many cities, and for short-term spatial change in any particular urban area. To begin to investigate questions of this nature one must know a great deal more about:

(1)   The detail of the urban built environment.
(2)   The dominant characteristics of urban manufacturing activity.

The built environment of most cities is very complex. Cities are not circular in shape and they are not composed of a series of concentric annuli of industrial property of different ages with the most modern in the suburbs. Industrial property for all but the smallest manufacturers is often confined to distinct areas of the city. Formerly free-standing towns of varying sizes have been engulfed by urban expansion resulting in pockets of older industrial property at varying distances from the historic urban core. Local authorities may rezone land into and out of industrial use depending on their needs and priorities. The demolition and clearing of obsolete industrial property − old mills and factories, railway sidings, gas works and docks − generates vacant land often zoned for industry in many different urban locations. And the building of urban motorways has brought direct access to the national highway system to the more central parts of the city. Thus, in the short to medium term the spatial change of manufacturing activity may not display a suburbanising trend.

In the generation of the component rates of change in urban areas small establishments and single plant firms are particularly important. Not only do establishments of this nature account for the majority of openings, closures and transfers in any city but their rates would appear to be higher than for other establishments. For example, Dennis (1978) discovered that in inner London in 1968 72 per cent of the plants fell into his smallest employment size category, 25–99 persons, with 74 per cent of plant closures between 1966 and 1974 also falling into this category. In outer London 56 per cent of the plants in 1968 and 61 per cent of the closures fell into this size category. In Clydeside in 1958 42.1 per cent of all plants employed between 6 and 24 employees yet over the following eleven years 50.7 per cent of plant openings were also in this size band (Cameron, 1973). In Greater Manchester between 1966 and 1975 Mason (1980b) discovered an inverse relationship between the transference rate and plant size, with approximately 50 per cent of all transfers possessing less than ten employees. The average size of all plants at the beginning of this study was 60.5 employees while those for the components of change were substantially less: openings 15.4, closures 35.0 and transfers 27.4. Indeed the evidence from most urban investigations would tend to suggest an inverse relationship between plant size and the component rates of change. Furthermore, since most small plants belong to single-plant firms these relationships also hold true for small firms. These results also indicate that small firms have very short lives. This is particularly so in inner city areas. For example in Greater Manchester in 1972 over

50 per cent of the plants in the inner area had been in existence for less than 10 years, with the equivalent proportion for the remainder of the conurbation being 43 per cent (Lloyd and Mason, 1978).

Although in terms of manufacturing net output there has been a decline in the importance of small firms (Prais, 1976) there are still many good reasons and opportunities for their existence including the production of goods for which the skills of craftsmen are required, the use of batch-production methods in industries where demand specifications may change rapidly, undertaking sub-contract work especially of a specialist nature, and the provision of specialist services. Small firms are also important in industries in which the capital costs of entry are low, while a small minority may be associated with entrepreneurial innovations (Lloyd, 1980; Cross, 1981). Thus, it is possible for small single-plant enterprises to be coming into being at all times. These firms will seek small sized and cheap premises, and in their early stages an entrepreneur's home may be used. Industrial premises suitable for small businesses can be found in many urban locations, however, important concentrations may often be available in the older central areas where the subdivided nature and state of disrepair of much of the property may make it particularly cheap to rent (Medhurst and Lewis, 1969). As we have already noted the chances of survival of many of these new firms are not good. Many will be single-product or service firms in most capricious markets such as clothing. Some will close due to the movement or closure of larger establishments to which they sold most of their products. Others, and especially those in central urban locations, will close due to lease termination and compulsory purchase for redevelopment. However, the principal reason for the high closure rate of small firms is inept management. The founders of small manufacturing firms are usually skilled on the production side of business but have little marketing and financial acumen. Thus, their lack of specialist management staff may give them a far greater chance of closure than other firms.

Following the work of Hoover and Vernon (1959) in New York it has often been suggested that the central areas of cities act as incubators, seed-beds or nursery areas for new firms. In these areas new small firms benefit from the external economies of accessibility and proximity to the full range of urban functions as well as being able to obtain cheap premises. Furthermore, it has been argued that as these firms grow and expand they will demand larger premises in less central locations. Unfortunately, with the exceptions of New York (Leone and

Struyk, 1976) and Leicester (Fagg, 1980) little evidence has been found to support these views. One of the important reasons for this, as Fagg (1980) correctly points out, is the use of establishment rather than firm or enterprise based data. Nevertheless, investigations of the incubator area hypothesis have demonstrated that few plants transfer from central urban areas for growth reasons (Leone and Struyk, 1976; Lloyd and Mason, 1978; Cameron, 1980). However, this is not to suggest that growth is not one of the most important reasons for the generation of transfers. Just that even when very large growth rates are recorded by inner-city transfers they often represent only very small absolute increases — say from four to eight employees — and may only result in the exchange of a very small workshop for a slightly bigger one. Thus, no marked change in the *type* of property demanded will result and such plants will probably remain in central locations. There are also a number of other good reasons for the generation of intra-urban transfers including lease termination and the serving of a compulsory purchase order (CPO) in designated redevelopment areas. The possible negative impact of redevelopment schemes on inner-city manufacturing activity remains a hotly disputed issue (McKean, 1975; Bull, 1979 and 1981; Darley and Saunders, 1976; Chalkley, 1979). However, there is no doubt that this process has accelerated the locational change of particularly small manufacturing plants. For example in inner Manchester between 1966 and 1972 25 per cent of transfers were precipitated by CPOs (Lloyd and Mason, 1978).

In the above analysis the suburbanisation of manufacturing activity was linked with specific technological developments. However, not all industries have adopted these technologies with equal vigour. In industries such as printing and publishing, clothing manufacture, certain wood crafts, leather goods and instrument engineering batch production methods in small establishments, perhaps organised into vertically orientated production systems, are in some cases still a profitable technological form. Thus, central urban locations may be preferred resulting in little suburbanisation. The New York garment complex (Hall, 1959, Leone, 1971) is a good example of such an industry. Of course some parts of these industries have changed technologically and moved from central locations. Partly as a result, along with changes in consumer affluence and demands, the once large clusters of their establishments in inner-city areas — sometimes called industrial quarters (Wise, 1951) — are now but small remnants of their former size. The decline of the clothing industries in London (Hall, 1962) and Montreal (Steed, 1976) may be viewed in this way. Nevertheless there are still a substantial number of firms in these industries displaying a marked preference for central urban locations.

It may be possible to argue that some cities appear to represent a 'dual economy' with small indigenous plants in declining or static industries in the city core and with the periphery populated by large, multi-regional and multi-national plants predominantly in the growth sectors of the local economy. Such a state of affairs Firn (1976) has been viewed as being in part the natural result of urban dendrochronology. However, such a form of dendrochronology, if it is a useful metaphor, should not be viewed in terms of particular firms or industries (Keeble, 1978), but more in terms of a decline in the usefulness and profitability of the urban fabric: the size and layout of premises and the transportation systems in inner-city areas becoming increasingly obsolete as changes in marketing, production and communication methods alter the most profitable ways, and therefore locations, in which goods can be produced. The manufacture of commodities that were not so amenable to new production forms has tended to remain in the inner areas, while many of the latest products have never been manufactured in central urban locations. As these changes have taken place the successful firms have vacated their central city premises, realising their value or converting them to other uses, while many of the weaker firms have gone into liquidation. Both these processes have released central production space which has often filtered down to increasingly marginal and transient enterprises.

Intra-urban Spatial Variations in Unemployment. In general unemployment rates are higher in the central areas of large cities than in the suburbs. For the British case Danson *et al.* (1980) have shown that the greatest variation between these two areas was in London where the 1971 male unemployment rate for the inner area of 7.5 per cent was 3.6 times the outer area equivalent. Some central parts of cities have recorded extremely high rates. In the Small Heath district of inner Birmingham, for example, male unemployment in early 1973 was at least 20 per cent, whereas for the conurbation as a whole the rate was 5.0 per cent (Department of the Environment, 1977a). However, one must not assume that suburban areas possess only low rates. In many public housing estates on the outer edges of the major cities very high unemployment rates have been recorded. Perhaps the most extreme example comes from Belfast where in January 1972 the Ballymurphy kilometre square recorded a male unemployment rate of 37.7 per cent (Doherty, 1981). For the beginning of the 1970s this was a dramatically high rate, but for the beginning of the 1980s in the depths of recession such high levels of unemployment for public housing areas are all too common. For example, according to Faux (1982) present day male unemployment rates in some of the worst housing areas of Glasgow can reach 50 per cent!

Some authors have argued that when the composition of the urban labour force in terms of its skills, age, sex, ethnic status, etc. are taken into account, there is very little spatial variation in unemployment rates within cities (Metcalf and Richardson, 1976; Richardson, 1980). The areas of high unemployment exist therefore because they are inhabited by people with characteristics for which high unemployment rates exist nationally. In this regard lack of skills is probably the most important characteristic (Department of the Environment, 1977b; Thrift, 1979).

However, before accepting this argument completely a number of other points should be considered. First, skilled workers may be registered as unskilled either because status downgrading may increase their chances of finding a job, or because they possess obsolete skills as a consequence of the changing technological demands of industry (Thrift, 1979). Second, one must consider how the spatial concentrations of the unskilled have come about. Has the provision of educational services in some areas been particularly poor so that being born in them would have increased a person's chances of being unskilled? Furthermore, have the housing allocation policies of local authorities led to the concentration of the poorer families into particularly run-down and 'notorious' estates from which some employers are unwilling to hire labour (McGregor, 1977 and 1979)? If the answer to either of these questions is in the affirmative then clearly where a person was educated or where a person lives will affect his or her chances of being employed. Finally, the changing geography of employment opportunities and public housing estates in the post-war period has added another locational dimension to this issue by creating in some cities a serious spatial mismatch of work places and residences. In some cases the members of families 'decanted' to suburban housing areas find the costs of travelling by public transport to central city workplaces prohibitive, while for others living in the inner city reverse commuting has posed similar problems.

Some Policy Issues

In the early 1980s most western countries have been characterised by large numbers of unemployed people: three million in the UK, nine million in the USA and two million in both France and Germany. In most countries this problem is at its worst in the major urban centres in which spatial concentrations of large numbers of the unemployed can be found. During the last decade the western world has to a lesser or greater degree been in the grip of a recession and as a result high levels of unemployment are to be expected. But if and when an upturn in the economic cycle takes place will very much of the growth impulse be realised in either the major conurbations or their inner areas, and will many new jobs be created in these areas? Given that the above discussion has identified powerful forces in the economy for manu-facturers first to substitute capital for labour, and second to locate away from the inner areas of cities, it is difficult to be optimistic on this point. Furthermore, given the current necessity for the manufact-uring sectors of most nations to improve their productivity and efficiency in the face of severe international competition it is highly unlikely that in the short to medium term any attempt will be made to attract major new manufacturing projects into these areas by offering financial incentives.

In the UK two new strands of policy are being attempted to aid the urban economies. First, a specifically area based policy involving inner-city partnerships between government and local authorities to improve the urban infrastructure and provide industrial property (Thrift, 1979), and more recently the creation of enterprise zones in which a package of generous financial incentives and planning concessions is available (Department of the Environment, 1981). Second, the beginnings of a national policy to aid the small firm sector. Given the large numbers of small firms in urban areas such an initiative could clearly have a part to play in any future urban regeneration programme.

Both these sets of initiatives will clearly benefit the major cities. However, there are two important reasons why they must both be viewed as little more than palliatives. First, although small firms have been most successful at generating new jobs (Birch, 1979; Storey, 1980) they are by virtue of their short lives very good at losing them (Lloyd and Dicken, 1979; Fothergill and Gudgin, 1979b). For example, of the 4,483 establishments with less than 20 employees in Greater Manchester in 1966 only half survived to 1975 and only 40 grew to employ 50 or more people. The employment produced by the survivors was therefore more than offset by the closures (Mason, 1980a). Programmes designed to improve the management abilities and the financial positions of small and new firms will undoubtedly reduce the scale of this enterprise carnage, but no major employment upturn will result in the foreseeable future. Only in the long term could this sort of policy generate significant employment benefits (Storey, 1980). Second, no matter how successful the small firm sector may be in generating employment the labour shedding problem of the major urban areas is so large that it can only be expected to have a modest impact upon it. To return to the Greater Manchester example, between 1966 and 1975 the closure of eight large plants and the contraction of a further 23 resulted in a loss of 43,388 jobs. Meanwhile, only five plants with more than 1,000 employees increased in employment by adding 5,839 new jobs (Mason, 1980a). How can the small firm sector be expected to offset job losses of this magnitude?

Given that attracting major new branch plants to relatively central urban locations will be extremely difficult for space and cost reasons the authorities with responsibility for these areas should probably first of all attempt to look after the existing firms in these areas, and especially the large employers, by helping them to modernise and perhaps expand. While this could initially result in job losses as capital replaces labour, it would be a small price to pay compared with the loss of these plants altogether. At the same time such a policy could include the nurture of the firms and industries, identified above, which appear to prefer central urban locations. However, at best such policies would only help to prevent further job losses. They would not help to create many new jobs.

In the short to medium term the chances of creating many new jobs in manufacturing in any western country is extremely small, and it is most unlikely that the promised expansion in high technology industries will recruit very many of the urban unskilled. Thus if reducing urban unemployment is deemed to be an important policy objective other ways must be found. For the unskilled in particular expanding the construction industry is one possible avenue to follow. There are certainly many areas of cities where the buildings, houses and roads are in urgent need of repair and replacement. However, given the strict monetarist policies of some governments on both sides of the Atlantic it is unlikely that sufficient public funds would be released to have any impact on the number unemployed by these means. Expanding the service sector is another possible option. However, as Thrift (1979) has observed, the paucity of research in this field makes it impossible to say if this is possible to achieve. Unfortunately, evidence has already been presented to indicate that services may be following manufacturing away from major conurbations. In addition, some forecasters have predicted large numbers of redundancies in the office sector as new electronic office equipment comes into common usage. Thus, it is perhaps doubtful whether the service sector will be able to furnish many new jobs in the future. One final option would be to cut the labour supply by reducing overtime, the working week and the male retirement age, and perhaps by increasing the school leaving age. In the near future such a scheme would be impossible to put into practice for cost reasons. However, given the willingness of industry at the present time to shed labour one must ask if such a policy would create many new jobs especially in the large cities and their inner areas? Would it instead create more retired people living off very small incomes in areas where concentrations of this unfortunate group can already be found? Furthermore would those in work be willing to take a substantial cut in their wages for the sake of providing jobs for the unemployed? Clearly the possibility in the foreseeable future of any major reduction in the unacceptably high levels of unemployment in the major cities is extremely small.

References

Beesley, M. (1955) 'The Birth and Death of Industrial Establishments: Experience in the West Midlands Conurbation', *Journal of Industrial Economics, 4*, 45–61

Birch, D.L. (1979) 'The Job Generation Process', *MIT Program on Neighborhood and Regional Change*, Cambridge, Mass.

Bull, P.J. (1978) 'The Spatial Components of Intra-urban Manufacturing Change: Suburbanization in Clydeside 1958–1968', *Transactions, Institute of British Geographers, N.S.3*, 91–100

Bull, P.J. (1979) 'The Effects of Central Redevelopment Schemes on Inner-city Manufacturing Industry with Special Reference to Glasgow', *Environment and Planning A, 11*, 455–62

Bull, P.J. (1981) 'The Effects of Redevelopment Schemes on Inner-city Manufacturing Activity in Glasgow: Some Evidence from two Central Development Areas', *Environment and Planning, A 13*, 991–1000

Cameron, G.C. (1973) 'Intra-urban Location and the New Plant', *Papers of the Regional Science Association, 31*, 125–43

Cameron, G.C. (1980) 'The Inner City: a New Plant Incubator?' in Evans, A. and Eversley, D. (eds.) *The Inner City*, Heinemann, London

Chalkley, B. (1979) 'Redevelopment and the Small Firm: the Making of a Myth', *The Planner, 65*, 148–51

Chisholm, M. and Oeppen, J. (1973) *The Changing Pattern of Employment: Regional Specialisation and Industrial Localisation in Britain*, Croom Helm, London

Corkindale, J.T. (1976) 'Employment in Conurbations', Paper presented to the Inner City Employment Seminar organized by the Centre for Environmental Studies, London

Cowan, D.J. (1971) 'Dynamic Aspects of Urban Industrial Location', unpublished
    PhD thesis, Ohio State University
Cross, M. (1981) *New Firm Formation and Regional Development*, Gower,
    Farnborough, Hants
Damesick, P. (1979) 'Office Location and Planning in the Manchester
    Conurbation', *Town Planning Review*, *50*, 346–66
Danson, M.W., Lever, W.F. and Malcolm, J.F. (1980) 'The Inner-city Employment
    Problem in Great Britain, 1952–76: a Shift-share Approach', *Urban Studies*,
    17, 193–210
Darley, G. and Saunders, M. (1976) 'Conservation and Jobs: the SAVE Report',
    Built Environment Quarterly, *2*, 211–27
Dennis, R. (1978) 'The Decline of Manufacturing Employment in Greater
    London: 1966–74', *Urban Studies*, *15*, 63–73
Department of the Environment (1977a) *Inner Area Studies: Liverpool,
    Birmingham and Lambeth. Summaries of Consultants' Final Reports*, HMSO,
    London
Department of the Environment (1977b) *Change and Decay. Final Report of the
    Liverpool Inner Area Study*, HMSO, London
Department of the Environment (1981) *Enterprise Zones*, HMSO, London
Dicken, P. and Lloyd, P.E. (1979) 'The Corporate Dimension of Employment
    Change in the Inner City' in Jones, C. (ed.) *Urban Deprivation and the Inner
    City*, Croom Helm, London
Dicken, P. and Lloyd, P.E. (1981) *Modern Western Society: A Geographical
    Perspective on Work, Home and Well-being*, Harper and Row, London
Doherty, P. (1981) 'The Unemployed Population of Belfast' in Compton, P.A.
    (ed.) *The Contemporary Population of Northern Ireland and Population
    Related Issues*, Institute of Irish Studies, The Queen's University of Belfast,
    Belfast
Drewett, R., Goddard, J. and Spence, N. (1975) 'What's happening to British
    cities?', *Town and Country Planning*, *43*, 523–30
Drewett, R., Goddard, J. and Spence, N. (1976) 'What's happening in British
    cities?', *Town and Country Planning*, *44*, 14–24
Engels, F. (1969) *The Conditions of the Working Class in England*, Panther,
    London
Fagg, J.J. (1973) Spatial Changes in Manufacturing Employment in Greater
    Leicester, *East Midlands Geographer*, *5*, 400–15
Fagg, J.J. (1980) 'A Re-examination of the Incubator Hypothesis: A Case Study
    of Greater Leicester', *Urban Studies*, *17*, 35–44
Faux, R. (1982) 'Home is where the Hell is', *The Times*, London, 24th February
Firn, J.R. (1976) 'Economic Microdata Analysis and Urban-regional Change:
    the Experience of GURIE', in Swales, J.K. (ed.) *Establishment based research*,
    Urban and Regional Studies Discussion Papers, 22, University of Glasgow,
    Department of Social and Economic Research
Firn, J.R. and Hughes, J.T. (1974) 'Employment Growth and Decentralisation
    of Manufacturing Industry: Some Intriguing Paradoxes', Papers from the
    Urban Economics Conference 1973, vol. 2, *Centre for Environmental Studies
    Conference Papers*, *9*, 483–518
Firn, J.R. and Swales, J.K. (1978) 'The Formation of New Manufacturing
    Establishments in the Central Clydeside and West Midlands Conurbations
    1963–1972: A Comparative Analysis', *Regional Studies*, *12*, 199–213
Fothergill, S. and Gudgin, G. (1979a) 'Regional Employment Change: A
    Sub-regional Explanation', *Progress in Planning*, *12*, 155–219
Fothergill, S. and Gudgin, G. (1979b) 'The Job Generation Process in Britain',
    Centre for Environmental Studies Research Series, p. 32

Frost, M. and Spence, N. (1981) 'Unemployment, Structural Economic Change and Public Policy in British Regions', *Progress in Planning*, *16*, 1–103

Gad, G.H.K. (1979) 'Face-to-face Linkages and Office Decentralization Potentials: a Study of Toronto' in Daniels, P.W. (ed.) *Spatial Patterns of office growth and location*, Wiley, Chichester

Gillespie, A.E. and Owen, D. (1981) 'The Current Recession and the Process of Re-structuring in the British Space-economy', Paper presented at the 21st European Congress of the Regional Science Association, Barcelona

Goddard, J.B. and Thwaites, A.T. (1980) *Technological Change and the Inner City*, The Inner City in Context, Paper 4, SSRC, London

Gripaios, P. (1977a) 'The Closure of Firms in the Inner City: the South-east London Case 1970–5', *Regional Studies*, *11*, 1–6

Gripaios, P. (1977b) 'Industrial Decline in London: An Examination of its Causes', *Urban Studies*, *14*, 181–9

Gudgin, G. (1978) *Industrial Location Processes and Regional Employment Growth*, Saxon House, Farnborough, Hants

Hall, M. (1959) *Made in New York*, Harvard U.P., Cambridge, Mass.

Hall, P.G. (1962) *The Industries of London since 1861*, Hutchinson, London

Haren, C.C. and Holling, W. (1979) 'Industrial Development in Non-metropolitan America: A Locational Perspective' in Lonsdale, R.E. and Seyler, H.L. (eds.), *Nonmetropolitan Industrialization*, Wiley, London

Hoare, A.G. (1975) 'Linkage Flows, Locational Evaluation and Industrial Geography: A Case Study of Greater London', *Environment and Planning*, *A*, *7*, 41–57

Hoch, I. (1972) 'Income and City Size', *Urban Studies*, *9*, 299–328

Hoover, E.M. and Vernon, R. (1959) *Anatomy of a Metropolis*, Harvard U.P., Cambridge, Mass.

Johnson, B.L.C. (1958) 'The Distribution of Factory Population in the West Midlands Conurbation', *Transactions, Institute of British Geographers*, *25*, 209–23

Kain, J.F. (1968) 'The Distribution and Movement of Jobs and Industry' in Wilson, J.Q. (ed.) *The Metropolitan Enigma*, Harvard U.P., Cambridge, Mass.

Keeble, D. (1976) *Industrial Location and Planning in the United Kingdom*, Methuen, London

Keeble, D. (1978) 'Industrial Decline in the Inner City and Conurbation', *Transactions, Institute of British Geographers N.S. 3*, 101–14

Keeble, D. (1980) 'Industrial Decline, Regional Policy and the Urban-rural Manufacturing Shift in the United Kingdom', *Environment and Planning A 12*, 945–62

Kemper, P. (1973) 'The Locational Decisions of Manufacturing Firms Within the New York Metropolitan Area', unpublished PhD thesis, University of Yale

Leone, R.A. (1971) 'Location of Manufacturing in the New York Metropolitan Area', unpublished PhD thesis, University of Yale

Leone, R.A. and Struyk, R. (1976) 'The Incubator Hypothesis: Evidence from five SMSAs', *Urban Studies*, *13*, 325–31

Lever, W.F. (1974) 'Manufacturing Linkages and the Search for Suppliers and Markets' in Hamilton, F.E.I. (ed.) *Spatial Perspectives on Industrial Organization and Decision-making*, Wiley, London

Lever, W.F. (1982) 'Urban Scale as a Determinant of Employment Growth or Decline'. Forthcoming in Collins, L. (ed.) *Industrial decline and regeneration*, proceedings of the 1981 Anglo-Canadian symposium. Department of Geography and Centre of Canadian Studies Publications, University of Edinburgh, Edinburgh

Lloyd, P.E. (1965) 'Industrial Change in the Merseyside Development Area,

1949-59', *Town Planning Review*, *35*, 285-98

Lloyd, P.E. (1970) 'The Impact of Development Area Policies on Merseyside 1949-67' in Lawton, R. and Cunningham, C. (eds.) *Merseyside: A Social and Economic Survey*, Longman, London

Lloyd, P.E. (1979) 'The Components of Industrial Change for Merseyside Inner Area: 1966-75', *Urban Studies*, *16*, 45-60

Lloyd, P.E. (1980) 'New Manufacturing Enterprises in Greater Manchester and Merseyside', *North West Industry Research Unit Working Paper 10*, School of Geography, University of Manchester

Lloyd, P.E. and Dicken, P. (1979) 'New firms, Small Firms and Job Generation: The Experience of Manchester and Merseyside 1966-75', *North West Industry Research Unit Working Paper 9*, School of Geography, University of Manchester

Lloyd, P.E. and Mason, C.M. (1978) 'Manufacturing in the Inner City: A Case Study of Greater Manchester', *Transactions, Institute of British Geographers N.S. 3*, 66-90

Lloyd, P.E. and Reeve, D.E. (1981) 'Recession, Restructuring and Location: A Study of Employment Trends in North West England, 1971-77', *North West Industry Research Unit Working Paper 11*, School of Geography, University of Manchester

Mason, C.M. (1980a) 'Industrial Decline in Greater Manchester 1966-75: a Components of Change Approach', *Urban Studies*, *17*, 173-84

Mason, C.M. (1980b) 'Intra-urban plant relocation: A Case Study of Greater Manchester', *Regional Studies*, *14*, 267-83

Mason, C.M. (1981) 'Manufacturing Decentralization: Some Evidence from Greater Manchester', *Environment and Planning A 13*, 869-84

Massey, D.M. and Meegan, R.A. (1978) 'Industrial Restructuring versus the Cities', *Urban Studies*, *15*, 273-88

McGregor, A. (1977) 'Intra-urban Variations in Unemployment Duration: A Case Study', *Urban Studies*, *14*, 303-13

McGregor, A. (1979) 'Area Externalities and Urban Unemployment' in Jones, C. (ed.) *Urban Deprivation and the Inner City*, Croom Helm, London

McKean, R. (1975) 'The Impact of Comprehensive Development Area Policies on Industry in Glasgow', *Urban and Regional Studies Discussion Papers 15*, University of Glasgow Department of Social and Economic Research

Medhurst, F. and Lewis, J.P. (1969) *Urban Decay: An Analysis and a Policy*, Macmillan, London

Metcalf, D. and Richardson, R. (1976) 'Unemployment in London' in Worswick, G.D.N. (ed.) *The Concept and Measurement of Involuntary Unemployment*, Allen and Unwin, London

Moseley, M.J. and Sant, M. (1977) *Industrial Development in East Anglia*, Geo-Abstracts, Norwich

Norcliffe, G.B. (1975) 'A theory of manufacturing places' in Collins, L. and Walker, D.F. (eds.) *Locational Dynamics of Manufacturing Activity*, Wiley, London

Penrose, E.T. (1963) *The Theory of the Growth of the Firm*, Blackwell, Oxford

Prais, S.J. (1976) *The Evolution of Giant Firms in Britain*, NIESR, London

Richardson, R. (1980) 'Unemployment and the Labour Market' in Cameron, G.C. (ed.) *The Future of the British Conurbations*, Longman, London

Roberts, J.M. (1979) The Diversification and Decentralization of Industry in Northampton, *The East Midlands Geographer 7*, 100-12

Royal Town Planning Institute, (1978) *Land Values and Planning in the Inner Areas*, RTPI, London

Saussy, G.A. (1972) 'An Analysis of the Manufacturing Industry in the New

Orleans Metropolitan Area', unpublished PhD thesis, University of Yale
Schwartz, G.G. (1979) 'The Office Pattern in New York City, 1969–75' in
    Daniels, P.W. (ed.) *Spatial Patterns of Office Growth and Location*, Wiley,
    Chichester
Stedman Jones, G. (1971) *Outcast London*, Oxford U.P., London
Steed, G.P.F. (1976) 'Locational Factors and Dynamics of Montreal's Large
    Garment Complex', *Tijdschrift voor Economische en Sociale Geografie 67*,
    151–68
Storey, D.J. (1980) 'Small Firms and the Regional Problem', *The Banker*,
    November issue
Struyk, R.J. and James, F.J. (1975) *Intra-metropolitan Industrial Location*,
    Lexington Books, Lexington, Mass.
Stuart, A.W. (1968) 'The Suburbanization of Manufacturing in Small
    Metropolitan Areas: A Case Study of Roanoke, Virginia', *South-eastern
    Geographer*, *8*, 23–39
Swales, J.K. (1976, edn) 'Establishment based research', *Urban and Regional
    Studies Discussion Paper* 22, University of Glasgow Department of Social
    and Economic Research
Thrift, N. (1979) 'Unemployment in the Inner City: Urban Problem or
    Structural Imperative? A review of the British experience' in Herbert, D.T.
    and Johnston, R.J. (eds.) *Geography and the Urban Environment. Vol. 2,
    Progress in Research and Applications*, Wiley, Chichester
Tulpule, A.H. (1969) 'Dispersion of Industrial Employment in the Greater
    London Area', *Regional Studies*, *3*, 25–38
Vallis, E.A. (1972) Urban land and building prices 1892–1969, *Estates Gazette*,
    222, 1406–07
Whitelegg, J. (1976) 'Births and Deaths of Firms in the Inner City', *Urban Studies
    13*, 333–8
Wise, M.J. (1951) 'On the Evolution of the Jewellery and Gun Quarters in
    Birmingham', *Transactions and Papers, Institute of British Geographers* 1949,
    15, 59–72
Wood, P.A. (1974) 'Urban Manufacturing: A View from the Fringe' in Johnson,
    J.H. (ed.) *Suburban growth*, Wiley, London

3 CRIME AND DELINQUENCY

D.T. Herbert

Although the 'applied' tradition is well-established in geography and has found clear expression is urban studies, the focus upon social problems is a much more recent feature. This focus can be contexted in several of the major trends which have affected research and teaching in urban geography over the past decade. Of particular importance was the call for 'relevance' in research which began to appear in more strident form in the early 1970s and the relationship which this was seen to bear with the strong interest in the 'social dynamic' of urban life which had emerged by that time. Urban geographers had become pre-occupied in the 1960s with the study of residential patterns, social areas and typologies of neighbourhoods, there was a natural progression from these studies to research which focused upon 'problem' areas of various kinds in the urban mosaic, neighbourhoods which were marked by 'deviance' in its various forms. The emergence of the social indicators movement, of analyses of levels of living and the quality of life offered a further point of reference and methodologies developed for other general purposes could be readily adapted for more specific uses. Underpinning shifts of emphasis of this kind within urban geography was the more general situation of a greater concern with new paradigms as the dominance of spatial analysis began to fade and, as significantly, the closer alliance of human geography with the philosophies and methodologies of the social rather than the natural sciences became apparent. Changes of this kind gradually provided a context in which greater research interest by urban geographers in a range of social issues could develop.

Crime and delinquency are two forms of deviant behaviour which have been studied in a detailed and academic way for almost two centuries. They share with most other forms of deviance the conceptual and pragmatic problems associated with definition and data, but have some advantage in that an elaborate set of laws, enforced by a highly professional judicial system, exists to define what constitutes a criminal act. Delinquency, as opposed to crime, has no special status in terms of type of offence; it is distinguished by the age of the offender rather than by the nature of the act. Crime and delinquency, therefore, are clearly labelled as offences against society, departures from its specified norms which constitute deviant behaviour. As criminology, concerned with the analysis of crime and delinquency, has developed, it has of necessity assumed a strong multi-disciplinary nature. There are many facets of criminal behaviour which fall into the natural 'compartments' of interest of a variety of academic disciplines which include law, psychology, sociology, and many others. Yet until the last decade of these two centuries of criminological study, there has been very little direct involvement by professional geographers. The awareness of spatial patterns of crime incidence and regional contrasts in types of crime was clearly present but as late as 1970, Scott (1972) could still describe his own interest in crime and delinquency as evidence of 'geographical deviance'. By the 1980s it can be fairly claimed that the study of social problems constitutes a main concern of urban geography and that crime and delinquency are prominent examples of such problems, but the concern is recent and the research is still, in many ways, in its formative stages.

In this review of the way in which the study of crime and delinquency is accommodated within modern urban geography a number of broad objectives will be followed. Firstly, the outlines from which a geographical perspective on crime and delinquency was to emerge can be traced from the longer history of criminology and the spatial facts which gradually and indirectly came from that research record. Secondly, some of the very significant issues regarding data and definition in any kind of criminological research will be discussed and the broad theoretical and methodological shifts of recent years will be recognised. Thirdly, the various ways in which geographers have begun to study crime and delinquency over the past decade will be identified and assessed.

The Origins of a Geography of Crime

Both Phillips (1972) and Sutherland and Cressey (1970) have referred to a 'cartographic' school of criminology which existed from 1830 to 1880, being initiated in France and spreading to England and other European countries. New sources of official data on crime rates and their variation over time and by regions provided a powerful stimulus to research of this kind. The adjective 'cartographic' arises from the frequent use of maps to show regional variations in crime rates and particularly urban-rural differences. Ecological association formed part of the analysis in the sense that the relationships of crime rates with other indicators of social condition were measured and discussed. Guerry (1833) used a series of annual reports to analyse French crime patterns and found consistent regional differences modified by factors such as seasonality. For example, crimes against property had a higher incidence in the north during winter, whereas crimes against persons reached highest levels in the south during summer. This particular finding pointed towards some climatic factor of relevance but Guerry also examined the effects of population density and education. Here the incidence of property crimes seemed related to urban areas, to wealth and to highest levels of literacy. Other writers have returned to the theme of natural environmental — particularly climatic — influences on criminal behaviour but there is no conclusive evidence of causal links (for a review of these studies, see Harries, 1980).

Fletcher (1849) provided a similar analysis of English crime statistics. He again found high crime rates in generally wealthier counties but recognised these as 'collecting' rather than 'breeding' areas for criminals, showing some awareness of the factor of opportunities. Fletcher's finding that high crime rates typified agricultural rather than manufacturing counties has proved eccentric in the context of this type of literature. H. Mayhew (1862), for example, showed variations by population density and 14 measures of deviant behaviour over the counties of England and Wales, and concluded that crime was most frequent in areas of industrial and urban character. Scott (1972) summarises findings from several parts of the world which show a link between urbanisation and crime and this remains a persistent feature at this scale of analysis; as societies move from a rural to urban character and particularly as large cities begin to emerge, official crime rates will rise.

Whereas the *cartographic school* was mainly concerned with regional patterns of crime, other contemporary accounts focus on the crime areas which had emerged within individual cities. These so-called 'rookeries' were found, for example, in most large British cities during the first half of the nineteenth century. Descriptions such as those of the St Giles district in London were commonplace:

The nucleus of crime in St Giles consists of about six streets, riddled with courts, alleys, passages and dark entries, all leading to rooms and smaller tenements . . . the lowest grade of thieves and dissolute people live in the immediate neighbourhood of the station house (Tobias, 1967, p. 131).

Elsewhere, Tobias identified particular streets, such as the appropriately named Twisters Lane, in which nine-tenths of the population
was estimated to be illegally employed. Districts such as these were
criminal quarters in a total sense; the law had little authority within
them and strangers were relucant to enter. Their context was a rapidly
emerging industrial society in which the physical form of urbanisation
was running ahead of an adequate system for the distribution of
resources (Stedman-Jones, 1971). The worst areas disappeared during
the latter part of the nineteenth century as the reform movement,
aimed at improving the living conditions, gathered force, with its assault
on overcrowding and public health; changes in the size and efficiency
of the police force were also relevant. Clearly the timing and extent of
the disappearance of the rookeries varied considerably fron one city
to another; in some cities vestiges of the old order persisted into the
twentieth century.

Mayhew (1862) provided more systematic studies of intra-urban
variations in crime rates. Using statistics for each of the seven police
districts in London in the mid-nineteenth century, he showed that
two of these distincts contained two-thirds of known criminals. Other
information on average income, tax and poor law assessments allowed
him to classify the districts and he was also able to disaggregate the
offenders and show clusters of particular types of crime. Throughout
the nineteenth century, therefore, a general geography of crime in
European countries was known. Regional variations had been identified
and measured, crime areas within cities were part of the common
knowledge of magistracy and police, and indeed most of the general
public.

The spatial ecology of crime was developed most fully in the 1920s
and 1930s with the work in America of Shaw and McKay (1942); their
best known study contained delinquency data on Chicago, collected
as part of the Illinois crime survey, and demonstrated a set of
techniques which they subsequently extended to other American
cities. For the Chicago area they mapped the homes of juvenile
offenders brought before the Cook County court at various periods
during the first half of the twentieth century. Analysis of these was
set in a spatial framework of square-mile grids, concentric zones, and
seventy-five designated community areas. Area rates were based on
appropriate populations-at-risk and a number of spatial generalisations
were identified. A range of cartographic procedures was used, including
dot maps to show actual distributions and rate maps to show areal
variations, and the generalisations focused on the observation of regular

changes from centre to periphery. The four main forms of deviance recorded — juvenile delinquency, adult crime, recidivism and truancy — were closely interrelated. Truancy, regarded as a good indicator of potential delinquency, was most typically found in areas adjacent to Chicago's central business district and inner city industrial areas, as was delinquency itself. Adult offenders were more likely to come from rooming house districts of the zone in transition. Overall rates depicted by gradients and zones revealed the regular progression from centre to periphery. Replication of the procedure for nineteen other American cities confirmed the spatial model, though two (apparently explicable) deviant sets of results were found in Baltimore and Omaha. A revised version of the original study (Shaw and McKay, 1969) added sets from the 1960s which showed a persistence of the general pattern, though there was an outward migration of crime rates in one sector, paralleling movements in the black district.

These areal generalisations were complemented by *ecological analyses* which correlated variables such as substandard housing, poverty, foreign-born population, and mobility, with high delinquency rates. The broad contrast observed was between central districts of poverty and physical deterioration, whose inhabitants were transient and possessed confused cultural standards, and the more stable family suburbs in which delinquency was invariably low. As Shaw and McKay conceived a delinquency area, it emerged from the transmission of delinquent behaviour to a point at which it dominated the attitudes and behaviour of the majority resident in a particular district. In addition to neighbourhood, other social group influences, such as those of family, school and play-group, were recognised and detailed individual case histories of known offenders (Shaw, 1930) were used to supplement more aggregate analyses. The work in Chicago in the interwar period established an ecological tradition in criminology and also provided a significant perspective to which geographies of crime could be related.

Schmid's (1960) study of Seattle provided one of the most comprehensive spatial ecologies of an American city since the pioneering efforts of Shaw and McKay. Data input for this study comprised 35,000 offences known to the police and 30,000 arrests over the time periods 1949 and 1950/51. This data set was reduced to 20 crime variables to which was added a set of 18 census variables calculated for each of 93 census tracts. Both offence and offender data were used in the analysis, with the justification that to define crime or delinquency areas it is important to know not only where offenders live but also where crime is committed. Schmid's summary statement on the geography of crime in Seattle was that the central segment of the city contained 15.5 per cent of the total population, 47 per cent of offences known to police and 63 per cent of the arrests. Areal generalisations by zones and gradients replicated the earlier model. From a factor analysis the three leading dimensions were successively described as low social cohesion and low family status, low social cohesion and low occupational status, and low family and economic status. Of these, the third dimension, with especially high loadings on unmarried and unemployed males and with the range of crime variables was described as the crime dimension *par excellence*. Scores from this factor located crime areas in Seattle:

> Urban crime areas, including areas where criminals reside and areas where crimes are committed, are generally characterised by all or most of the following factors: low social cohesion, weak family life, low socio-economic status, physical deterioration, high rates of population mobility and personal disorganisation. (Schmid, 1960, p. 678)

This type of areal and ecological generalisation has been replicated in studies of other parts of the world. A succession of studies of Liverpool (Jones, 1934; Castle and Gittus, 1957) revealed clusters of social defects, including crime in the inner areas of the city which contained high levels of immigrant population and overcrowding. Studies in Third World countries are rare and somewhat inconsistent. Caplow (1949) found crime gradients to be reversed in Guatemala City with highest rates on the urban periphery; Hayner (1946) found a closer resemblance to the American model in Mexico City though some peripheral clusters existed which corresponded with shanty towns. Any attempt to draw generalisations from studies of Third World cities is fraught with difficulty. The few available studies are based on small sample sizes and large observational units: the adequacy of their data bases is extremely doubtful. Differences in areal patterns from Western countries reflect broader contrasts in the social geography of the city whereas general ecological correspondence with poverty and disadvantage is more consistent, though at different scales.

These early studies established the spatial ecology of crime and delinquency and provided a base for the geographical perspectives of the 1970s. As geographers began to study crime, however, they were faced with new kinds of data problems.

Sources of Crime Data

At an aggregate scale there is a considerable supply of statistical data relating to crime and delinquency in most Western societies. In the United Kingdom, the Home Office publishes its detailed *Criminal Statistics* on an annual basis and, within the limits of the aggregations used, these show variations in offence rates over time, comparisons between types of offences, and interregional differences. In addition to information on offences by police districts, the *Criminal Statistics* contain data on sentencing patterns at the various levels of courts, both crown and magistrates' courts, by police district and crown court centre. Information on applications for legal aid are provided for petty sessional division. This last category of information is composed of the largest number of territorial units but for the largest sets of data the police districts, of which there are only 43 in England and Wales, form the standard geographical framework. For the United States, the annual *Uniform Crime Reports* contain both a large amount of data and a considerable range of analyses and comparable publications exist in other countries of the developed world.

Over time there have been technical changes in the presentation of data which make comparative analysis more difficult. Some of these involve definitions of offences and changing forms of punishment; others, more directly relevant to spatial analyses, involve adjustments to boundaries of police force areas which form the basic data recording units. In the mid-1960s, for example, there were major reappraisals, affecting both the general form and content of the *Criminal Statistics*, accompanying an amalgamation of police forces in England and Wales from the previous 125 to 43. Besides reducing the amount of detail possible in terms of geographical variation, these new police force areas maintain the lack of correspondence with other administrative districts which has been a consistent feature of the data source. Official crime rates are normally calculated against base populations of 100,000 but offence-specific rates which use more relevant base populations (see Boggs, 1966) are more telling measures. The logic of this argument is that crime (offence) rates should be calculated against 'units at risk'; shoplifting, for example, should be expressed as a ratio of number of shops and residential burglaries against numbers of residential properties. For these purposes Baldwin and Bottoms (1976) adopted an index of industrial and commercial properties, Harries (1980) suggested a 'risk-related' crime rate formula, Bradbury (1982) experimented with a number of indices which expressed the frequency of offences as a ratio of land area and added weightings according to the gravity of the offence.

Despite this apparent surfeit of statistical bases from which to develop analyses, few data sources are regarded with more scepticism and mistrust. Sources of dissatisfaction include collection and classification procedures, definitions, the nature of crime and the motivations of attempts to record it. These issues have been discussed in serveral major texts; see, for example, Carr-Hill and Sterns (1979), Hindess (1973), and Kitsuse and Cicourel (1963).

Criminological Theories

Although it can sensibly be argued that all criminological theories are relevant to a geography of crime, there is nevertheless some virtue in selecting a smaller number of concepts and theories for discussion. Some of these concepts can be discerned in very early studies but again it is in the spatial ecology of Shaw and McKay (1942) that they were placed in systematic order. Shaw and McKay recognised indices of mobility, housing quality, poverty and ethnic status as correlates of high delinquency rates in Chicago but did not seek to suggest they were directly causal. Rather they argued that these, along with delinquency rates themselves, were symptoms of some underlying condition from which deviant behaviour might emerge. The theory of social disorganisation was developed from this basis and suggested that in the absence of a stable form of society with legalistically based codes of behaviour and established norms and values, precipitating conditions for criminality would exist. Although it is evident that high delinquency rates occur in some stable, 'organised' inner-city districts, the concept continues to receive attention (Baldwin and Bottoms, 1976). A related and influential concept is that of anomie, originally associated with Durkheim and the notion of normlessness, which has also been used in attempts to explain criminal behaviour. Whereas it could be argued that both anomie and social disorganisation could be linked to territories within large cities, neither was developed primarily in relation to local environmental circumstances. Both sought to relate an individual's disposition towards criminality to underlying *structural* conditions; it was the nature of the encompassing societal system and the individual's place within it which formed the main point of reference.

Shaw and McKay were also associated with the theory of cultural transmission in which the delinquent tradition, 'nurtured' among some sections of society, could again be viewed in a territorial context. H. Mayhew's (1862) description of rookeries in London, where children were born and bred to the business of crime, is closely related to this notion of delinquent behaviour being transmitted over time and space. Sutherland's (1940) theory of differential association, which suggests that a person becomes delinquent because of an excess of definitions favourable to violations of law over definitions unfavourable to violations of law, also has common features. The steps from theories of this kind, which tend to view the offender in local circumstances, to the subcultural theories which dominated a couple of decades from the later 1950s, were short. Taylor (1973) regarded the subculturalists as inheritors of the Chicago tradition and Phillipson (1971) suggested that theories such as anomie, social disorganisation and cultural transmission, all provided the basis for a subcultural approach.

A subculture suggests the existence of identifiable groups within which particular sets of knowledge, beliefs, values and normative codes of behaviour are typical. The concept of a delinquent subculture arises from the fact that some of the elements associated with such a group may be illegal and at odds with those adopted by the wider society. Matza (1964) described subcultural theory as a 'modern rendition of positivism' in which peer groups form reference groups establishing behavioural norms. There are variants on the theory of subculture. Miller (1958) generalised from his analyses of lower-class youths and saw delinquency as part of their traditional behaviour or one of their 'focal concerns'. Cloward and Ohlin (1960) saw subcultural delinquency as a reaction by working-class youths to their failure to achieve in middle-class terms; delinquency was not a reaction to middle-class standards but a refusal to legitimise them. For Cloward and Ohlin, the content of a delinquent subculture was significantly shaped by its local milieu, and they classified the various responses in that milieu as criminal, conflict and retreatist. Matza (1964) emphasised the extent to which subcultures could be integrated with the surrounding social world; delinquents are not isolated but are encircled by institutions which uphold conventional values. For Matza most young offenders retain choice, and 'drift' between conventional and delinquent behaviour; they rationalise deviance in terms unacceptable to society but are capable of reform.

Subcultural theory has particular interest for social geographers as local environment can be abstracted as one framework for behaviour. The extent to which groups have been linked with territories within the city is variable, but for most theorists these were not central features. Links with more closely researched themes by geographers occur with concepts such as neighbourhood and 'community' within which localised sets of values and norms can be hypothesised. Subcultural theory has some attraction, therefore, in that it can be theorised in spatial terms, groups can be identified with places. Subculture as a concept also has more flexibility than many other positivist stances. It can accommodate the fact that many 'non-delinquents' are found in delinquency areas (Cloward and Ohlin, 1960), it includes the function of free will, drift and choice (Matza, 1964) and it provides many attractions as a working hypothesis for geographers focusing upon the intervening effects of local environments.

The concepts which have been briefly described have their roots in the long-established tradition of criminological research. They have attraction to geographers seeking to study criminal behaviour in that they proceed from familiar positivist stances and have a focus of empirical research which can, without too much difficulty, be related to the local environments in which offenders live and in which offences occur. There have been changes in criminological theories in the 1970s however, highly critical of conventional methodology, which have close parallels with changing approaches in urban geography. Sociological theories of deviance provided emphatic moves away from the positivist and correctional stances of traditional perspectives arguing (Rock, 1973) that conventional concerns with the pathology of the individual offender, or with the groups in which he organised his behaviour, were misplaced. An early 'replacement' by labelling theory and symbolic interactionism was itself superceded by a new radical criminology which sees both criminal law and behaviour as outcomes of political power relationships which emanate from the process of production. The similarities between these trends and those in human geography, which led to the emergence of behavioural and humanistic perspectives on the one hand and radical geography on the other, are strong. The emergence of competing and often apparently contradictory perspectives in both criminology and geography does pose problems. Figure 3.1, which has been argued in more detail elsewhere (Herbert, 1982), offers one framework in which various geographical perspectives on crime can be accommodated. Level 1 offers a 'niche' for a radical geography which suggests that crime emanates from societal structures which produce inequalities, 'have-nots', and the frustrations which lead to legally-defined crime. Level 2 accommodates the useful managerial perspective which to date is minimally studied by geographers (see,

Figure 3.1: A Conceptual Model for a Geography of Crime

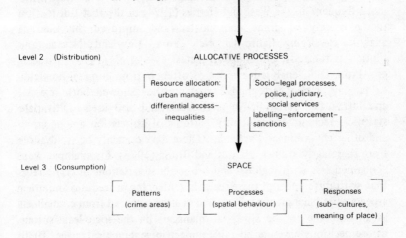

however, Harries and Brunn, 1978; Davidson, 1981) and could incorporate some aspects of labelling theory. Level 3 provides a context for the most traditional geographical approaches of mapping the incidence of crime in space, for a more behavioural analysis of spatial processes, and for a humanistic perspective on the values and meanings attached to space and place. It is at this level, at the scale of outcomes in local space, that the bulk of research into the geography of crime has been placed. The discussion of this research, which follows, will reflect these emphases but will also seek to show how links with the other 'levels of analysis' occur.

Modern Geographical Approaches to Crime and Delinquency

As geographers in the 1970s began to research into crime and delinquency in specialised and systematic ways, two broad influences seemed to direct their approaches. Firstly, there were incentives to analyse in a more detailed manner, some of the observed spatial regularities which had led to descriptions of the regional variations of crime rates and, secondly, there were temptations to apply a range of newly acquired analytical techniques to a new form of data. Both of these influences tended to push the new geographers of crime into a type of research which had strong descriptive and empirical qualities. In the United States, for example, early writers like Lottier (1938) had observed the fact that highest murder rates occurred in South-eastern states, that robberies were highest on an axis from Washington to Texas and larceny was more frequent in the West and Harries (1974) in the first full text on the geography of crime consolidated and improved this kind of analysis classifying California, New York, Maryland, Nevada and Florida as the most criminogenic states. Pyle *et al.* (1974) in studies of Akron, Ohio analysed a series of offences from criminal homicide to larceny. Using cartographic analyses they showed both regional and intra-urban variations in crime rates and also used multivariate statistical techniques to identify patterns of association among crime and other environmental variables. There have certainly been advances from these early studies. Harries and Brunn (1978), for example, were concerned less with regional and sub-regional variations in crime rates than with differences at these scales in sentencing procedures and other aspects of judicial practice. They had in a sense moved from crime itself as an outcome to the study of 'managers' in the socio-legal system whose decisions have considerable impact in geographical space. There are also moves from descriptions of regional variations towards a framework in which some closer approach to explanation was possible. Southern violence in the United States, for example, is related (Harries, 1980) to a set of ideas based on a subculture of traditions, cultural development, and contemporary life-style in particular states. Rengert (1980) and others have sought to form theoretical frameworks from which geographies of crime may develop.

There have been some continuing interest in inter-urban crime rates. Haynes (1973) offered the concept of density of opportunity and found it useful in an analysis of 86 American cities; Harries (1980) tested this model and found it of limited value. More recently, Brantingham and Brantingham (1980) have advanced a different kind of explanation by attempting to relate variations in inter-metropolitan crime rate to measures of the employment and occupation mixture of urban populations. Although this kind of research, at regional and inter-urban scales, continues to have valid contributions to make, it is at an intra-urban scale — within the city — that most geographers have focused their research effort.

Developments from the Ecological Tradition

The spatial ecology approach developed most comprehensively by Shaw and McKay (1942) has continued to influence much empirical research. Most commonly these studies show areal patterns and ecological associations at an aggregate scale of analysis using small area statistical bases. Areal pattern studies can be used to test the existence of zonal gradations in crime rates from city centre to periphery. Pyle *et al*. (1974), for example, used a number of cartographic techniques to depict crime distributions in Akron, Ohio; Corsi and Harvey (1975) showed crime surfaces in Cleveland. Studies of this type in North America tend to confirm the existence of zones and gradients and support Scott's (1972) earlier contention that these spatial generalisations have stood the test of time. Evidence from British cities is far less convincing and it is clear that as public sector intervention in the urban housing market has redistributed people from the inner city, it has also transferred some forms of deviance to peripheral estates. Morris (1957) found in Croydon that clusters of delinquency residence were related to public sector housing; Timms (1965) noted clusters in both central Luton and in some older municipal housing projects; similar patterns occur in Sheffield (Baldwin and Bottoms, 1976) and Cardiff (Herbert, 1976). Most areal studies have been confined to aggregate data but where, as in the Cardiff study, it was possible to examine patterns at an individual level these tend to show clusters in particular sections of larger neighbourhoods. Morris (1957) talked of 'black' and 'white' streets and other studies have described segments of estates where problem families appear to live. Research into ecological associations of high crime and delinquency rates also shows remarkable consistency in terms of its results. Most generally there is a strong association between crime and delinquency and what might be classed as 'poor environments'. Figure 3.2 summarises the main indicators which either singly or in various combinations tend to correspond with a high incidence of offenders.

Schmid (1960) found high proportions of males and unemployed to be the main indices of crime; a large number of studies has shown a consistent and inverse relationship between crime and delinquency and social class or socio-economic status, however measured (see Davidson, 1981). The Cardiff study (Herbert, 1976; Evans, 1980) found broken homes and large families to be significant features of delinquent groups; others have shown above-average offender rates among minority ethnic groups. Findings of these kinds have general descriptive value and

Figure 3.2: Common Objective Attributes of Known Offenders

category	indicator		sub-group at risk
DEMOGRAPHIC	age	⟶	young
	sex	⟶	male
	marital status	⟶	single
	ethnic status	⟶	minority group
	family status	⟶	broken home
	family size	⟶	large
SOCIO-ECONOMIC	income	⟶	low
	occupation	⟶	unskilled
	employment	⟶	unemployed
	education	⟶	low attainment
LIVING CONDITIONS	housing	⟶	substandard
	density	⟶	overcrovded
	tenure	⟶	rented
	permanence	⟶	low, transient

confirm what has been known for some considerable time, that officially defined offenders are drawn in disproportionately high numbers from low-income groups in the lower strata of society and from residential areas of cities where they live. Researchers have often moved from evidence of statistical association of this kind to theories of criminal causation but such 'leaps' are rarely justified. Lander's (1954) attempt to propose an anomie explanation for delinquency in Baltimore from this kind of analysis has been justly criticised as was Brown *et al.*'s (1972) support for anomie, sub-culture, differential association, retreatism and double-failure hypotheses on the basis of a cluster analysis of 35 variables. Hirschi and Selvin's (1967) plea for procedural vigour and circumspection in inferential analysis deserves continued emphasis.

Crime and Delinquency Areas

The areal analyses which identify clusters of offenders in space and the ecological analyses which demonstrate their environmental correlates are brought together in the idea of crime and delinquency areas. Most references to such areas in the criminological literature refer to districts within cities which have disproportionate shares of both offences and offenders. The definition is stated in relative terms but the actual nature of such areas will be highly variable over place and time. Defined in these empirical terms, crime areas were identified in mid-nineteenth-century British cities (Tobias, 1967); in American cities in the early twentieth century (Shaw and McKay, 1942); and are observed in various forms in modern cities (Damer, 1974). Mack (1964) argued that although the mobility of the adult criminal makes the crime area a thing of the past, every large city contains delinquency areas characterised by exceptionally large numbers of young offenders. Such areas exist but are nothing like the equivalents of Mayhew's (1862) rookeries; Davidson (1981) was more disposed to argue for the persistence of crime areas and saw inner-city slums as residential environments of long-term recidivists.

Within the delinquency areas of modern cities, known offender rates never approach 100 per cent of the population at risk and are rarely a simple majority. Edwards (1973) calculated prevalence rates for offences ever committed among a cohort of boys in Newcastle upon Tyne and found the highest ward rate to be 54.2 per cent. Forman (1963) and Kobrin (1951) found highest area rates of 20 per cent for court cases and 30 per cent for police contacts. In a detailed study of a British city, Mack (1964) discovered a criminal residence rate of 32 per cent in his worst street. Official data understate the rates and hidden delinquency is probably high in these areas, but the concept of delinquency area need not imply that all the at-risk population will become delinquent. Place of residence is but one frame of reference for adolescent behaviour. There are others such as family, school and teacher, workplace and interest group which may modify its effects. More narrowly it is argued:

The reasons for not all working class children becoming delinquent may be listed as follows: (i) degree of stress resulting in psychiatric delinquency tends to vary with circumstances of individual families and personalities; (ii) not all delinquents . . . will commit acts which are specifically illegal; (iii) by no means all those who commit illegal acts will be detected and prosecuted and identified as delinquent within the definition of the law (Morris, 1957, p. 176).

Most modern descriptions of delinquency areas in the criminological literature have been conservative in their claims and have focused on their legibility within the city rather than upon any offender-producing roles which they might have. This is in contrast with nineteenth-century observers, who consistently proposed the 'breeding-ground' hypothesis (Mayhew, 1862). Shaw and McKay (1942) showed that delinquency areas could be identified, but stressed that the criteria used could not furnish explanations; these had to be sought in the field of more subtle human relationships and social values. Reservations of this kind are typical but more recent reviews continue to suggest that the significance of such areas cannot be overlooked:

When we talk of a criminal area or of a delinquent sub-culture we are not saying that every individual living spatially close to offenders is so powerfully conditioned by their attitudes and behaviour that he is obliged to break the law himself. What we are saying is that within a broad zone which can be drawn upon a map, a very substantial number of people commit offences and there is a general social tolerance extended towards this behaviour . . . the area as a whole is delinquency-producing . . . the exceptions do not disprove the generalisation (Mays, 1963, p. 219).

Whereas some criminologists (Mannheim, 1965) argue that subcultures have no necessary territorial base, much empirical evidence suggests that they do. Delinquency areas, once recognised, can be analysed in more detail. Shaw and his associates used such frameworks to examine the individual characteristics of offenders and to locate particular remedial schemes such as the Chicago Area Projects. More recent research has adopted similar attitudes and has proceeded from use of objective indicators to define areas with high offender rates and subsequently to the use of subjective data to investigate their character. Scott (1972) suggested that once the incidence of crime had been plotted the climate of opinion in the streets could be studied. Morris (1957) was especially interested in proceeding to an analysis of the common 'social universe' of which both offenders and non-offenders were members.

Herbert (1976) developed a geographical approach to the study of delinquency areas which used an area sampling framework to examine the existence of a 'neighbourhood effect' incorporating elements of an urban geographer's concern with territory and the sociological concept of sub-culture. Having mapped delinquency rates for small areas and classified these same units into social area types, sets of areas were chosen which were similar in a range of socio-economic and demographic characteristics but which differed markedly in terms of numbers of offenders. Testing parental attitudes towards education, punishment and their definition of what constituted serious mis-behaviour, there was a strong trend of evidence to show that values and attitudes were more positive and potentially helpful in those areas with low delinquency rates. Wilson (1980) provided some useful related evidence with her conclusion, from a differently constructed study, that parental supervision is the most important single factor in determining juvenile delinquency. There are two kinds of findings here. The first which is reasonably undisputed is that the subjective values and examples which surround children as they move through adolescence influence their behaviour; where such values are deleterious, they may 'promote' delinquency. The second is empirically evident but much more difficult to conceptualise, namely that such values are associated with particular 'territories' and these are likely to be delinquency areas. The empirical basis exists because such 'areas' are evident in larger cities; the conceptual difficulties arise because area *per se* has limited meaning independent of the social class or ethnicity or life-styles of the people who inhabit it and remains an 'aggregate' within which individual exceptions may well occur (Hamnett, 1979). Even with these caveats a 'neighbourhood effect' as an empirical reality in probablistic terms remains a concept worthy of exploration by geographers.

The Emergence of Problem Areas

The facts of deliquency areas can be established and their internal characteristics can be examined by detailed surveys; remaining questions concern the emergence and persistence of such areas. These questions are also difficult, as their resolution requires historical data of considerable quality and of a type which is rarely available for the inner-city problem areas which have attracted most attention. A more recent British phenomenon is the emergence of problem areas in peripheral urban locations which coincide in part or whole with large public sector housing projects, often of interwar construction. Baldwin (1975) suggests that in Sheffield the continuing problem of criminality remains, for the most part, in those estates built in the 1930s or earlier; These are the 'problem estates' and are regarded as such by local population, local agencies, social workers and police. The recency of these estates and the fact that they were constructed and documented within the public sector, has made more possible the detailed scrutiny of ways in which they have emerged.

Baldwin (1975) identified three types of explanation for the emergence of problem estates and assessed the merits of each in the light of the Sheffield study:

(1) those theories which focus on allegedly high turnover rates in delinquent estates;

(2) suggestions that a paucity of social and recreational facilities for youngsters may lead to problems;

(3) suggestions that urban managers have some role in creating problem estates by the criteria they use to allocate tenants to housing in various parts of the city.

Recent research has concentrated on the third type of explanation and has developed this in the more general context of labelling processes. Gill (1977) studying a problem housing project in Liverpool suggested that:

It was local planning and housing department policies that produced Luke Street. The action of the police and the stereotyping of Luke Street as a 'bad' area were crucial but secondary processes (p. 187).

The longest established argument in support of the view that housing departments create problem estates is that a 'dumping' policy ensures that tenants, preselected because of adverse qualities are gathered together on particular estates. Wilson (1980) argues that a self-selection process may occur which has the effect of bringing similar people together but Gill (1977) viewed this as a process by which those least able to compete for housing end up in the least desirable areas. There is evidence that early stigmatisation of estates, often associated with the first wave of tenants, tends to have prolonged effects on its future reputation and character; Figure 3.3 summarises these and other contributory factors.

Figure 3.3: Contributory Factors in the Emergence and Persistence of Problem Estates

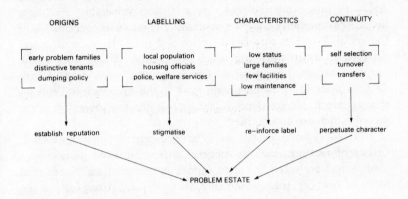

Offence Patterns and the Concept of 'Vulnerable Areas'

The empirical study of crime is approaching its two hundredth year. One of the most interesting current thrusts in criminology returns us to concerns which marked the early phases of the systematic study of crime, namely the distribution of crime in space and what that implies for crime prevention (Brantingham and Brantingham, 1975, pp. 11-12).

Although concern with the evidence of offences rather than offenders and with crime prevention rather than crime causation can be traced to the origins of criminology, the 1970s has witnessed a resurgence of interest in these themes. Such an approach has distinct advantages. Firstly, broad contexts can still be recognised but active research may focus on narrower and more immediate factors related to criminal activity. Secondly, these factors involve analyses of the local environments within which offenders, police, and victims interact. Thirdly, the fallibilities of official statistics are reduced to the extent that much more is known about offences than offenders. Fourthly, the context of the local environment allows access to the roles of victims and their reactions to crime or fear of it. Lastly, such studies enable a much closer focus upon preventive strategies and to policies which may protect those who live in 'vulnerable' areas.

There are two broad theories of central interest to geographers interested in the study of offence patterns. The first is concerned with the identification and characterisation of vulnerable areas or locations in the urban environment for particular types of crime. Such areas are often characterised by distinctive qualities of physical environments such as design, layout and accessibility; Newman (1972) with his thesis of defensible space was instrumental in rekindling debate on this issue and with promoting the idea that safer, less vulnerable, and more liveable environments could be designed. The second is concerned with the social dynamics attached to place and space. How do offenders perceive the environments in which they behave? Which 'cues' in the forms of targets, opportunities, familiar ground, perceptions of risk do they react to? How do victims react to the environmental circumstances which render them vulnerable? Is there a journey-to-crime pattern which can be identified?

Different types of offence clearly have contrasted spatial expressions. Crimes of violence, such as robbery, are concentrated at points of conflux and are often linked to various institutional and commercial forms of land-use; there is normally, for example, a strong focus in the central city. Residential crime, such as burglary, has a different spatial distribution and is primarily associated with particular types of residential areas. There have been numerous studies of the incidence of residential burglary, most of which have incorporated some elements of the defensible space idea. Waller and Okihiro (1978) examined burglary in Toronto and found that proximity of affluent housing to public sector projects was the best correlate of high burglary rates. They could not isolate a 'community' or social cohesion effect but attributed a good deal of significance to household security and patterns of dwelling occupance. Davidson (1981) found that burglary in Christchurch, New Zealand was overwhelmingly concentrated in the least affluent areas of the city; the poorest quarter of the city's census tracts suffered 66.4 burglaries per 1,000 households compared to the richest quarter's 12.8. Burglary in Christchurch was described as a highly opportunist criminal activity closely tied to neighbourhoods where offenders reside. Repetto (1974) found highest burglary rates in central residential districts of Boston but attributed strong significance to social cohesion within neighbourhoods. Herbert (1982) observed similar characteristics in parts of inner Swansea in which highest burglary rates seemed to coincide with mixed neighbourhoods of low social cohesion in which numbers of sub-divided dwellings were high.

Whereas any set of offenders will include a highly professional group who plan and execute their offences in systematic ways, there are many indications that much crime is opportunist and responds to perceived situations in local environments.

Stimulus conditions, including opportunity for action presented by the immediate environment are seen to provide, in a variety of ways, the inducements for criminaltiy. These are modified by the perceived risks involved in committing a criminal act; the anticipated consequences of doing so; and – in a complex, interrelated way – the individual's past experience of stimulus conditions and of the rewards and costs involved (Mayhew *et al.*, 1976, pp. 2-3).

Figure 3.4A attempts to codify elements of the situational context within which offences occur. Offence location can be precisely identified and a set of environmental features related to that location can be measured. A comparative analysis may involve offence locations (based on official statistics) and non-offence locations, though identification of the latter is a research problem. Other elements related to offence locations can be contexted in ideas of defensible space, labelling theory, and recent work in social geography. *Opportunity* for an offence is a basic stimulus, and this is likely to be affected by the design qualities of the adjacent built environment. *Surveillance* is a significant aspect of design through qualities such as observability and nature of access. *Social* factors modify design qualities and physical space in many ways. Attitudes towards a dwelling in terms of security and behaviour in terms of frequency of occupance are critical. In the neighbourhood, the form of proprietary control over communal space and attitudes towards local environments outside the immediate dwelling are as significant.

Additional to environmental 'cues' for residential crime, are the *area* features which are relevant and Figure 3.4B shows some of the related hypotheses. These area hypotheses start from the assumption, well attested in the social geography of the city, that the urban area forms a mosaic of 'neighbourhoods' which can be defined and delimited. Some of these neighbourhoods, or parts of them, may have qualities which make them collectively more vulnerable to offenders. The *'border-zone' hypothesis* was suggested by Brantingham and Brantingham (1975) and proposes that where well-defined neighbourhoods exist, their peripheral or border components will have greater vulnerability to crime; empirical evidence from Talahassee provided support for this hypothesis. The logic of the hypothesis is that offenders will see most attractive targets on the edges of their own neighbourhood, within which they have relative anonymity; borders of greatest vulnerability are likely to be those adjacent to high crime-rate areas. The *variability hypothesis* has been emphasised by Winchester (1978)

Figure 3.4: Understanding Offence Patterns

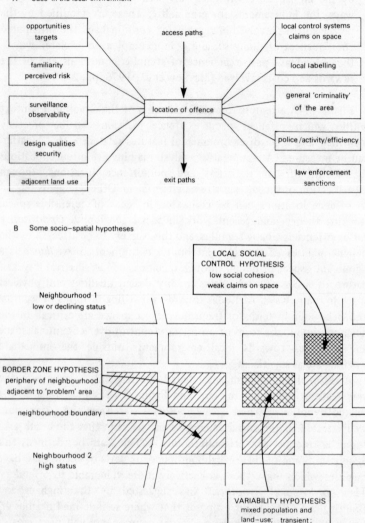

A   'Cues' in the local environment

- opportunities targets
- familiarity perceived risk
- surveillance observability
- design qualities security
- adjacent land use

access paths

location of offence

exit paths

- local control systems claims on space
- local labelling
- general 'criminality' of the area
- police/activity/efficiency
- law enforcement sanctions

B   Some socio-spatial hypotheses

LOCAL SOCIAL CONTROL HYPOTHESIS
low social cohesion
weak claims on space

Neighbourhood 1
low or declining status

BORDER ZONE HYPOTHESIS
periphery of neighbourhood
adjacent to 'problem' area

neighbourhood boundary

Neighbourhood 2
high status

VARIABILITY HYPOTHESIS
mixed population and
land-use; transient;
subdivided property

who interprets it in various ways but, relevant to the present context, argues that variability of conditions *within* areas may present many opportunities for crime. Areas in some form of transition in which, for example, a minority higher social class group may exist as a residual of former higher status or as a forerunner of upgrading, may have higher offence rates than stable, more uniform areas. As a hypothesis, this is useful but needs refinement in terms of both measurement and theory. What dimensions of variability for example are most critical? Tenure, social class, demographic structure, mobility are all potentially relevant. Winchester stresses variability of opportunities but other factors may be of significance (see Smith, 1981). As with this hypothesis, the third which may be termed *local social control* is not single-stranded. The social cohesion idea is relevant, as long-established, well-acquainted, neighbourly sets of people are less likely to be affected by offences (Herbert, 1982); parts of the urban environment which are lacunae or vacua between local control systems are especially vulnerable (Ley and Cybriwsky, 1974); target areas may be typified by low-risk, ease of access, and avoidance from observation (Davidson, 1981). All of these scenarios can be accommodated within a local social control hypothesis. Research into local environment and offence patterns offers useful lines of development in which environment/behaviour relationships are relatively unambiguous, data are more available and findings more directly translated into crime prevention policies. These kinds of policies, aimed at target-hardening or security-improvement in particular locations are criticised in the sense that they merely *displace* offences to less well guarded locations. There is good evidence, however, (Mayhew *et al.*, 1976) that opportunist offenders are deterred and not merely displaced by such policies.

Spatial Behaviour of Offenders

The spatial behaviour of offenders is one obvious line of study by geographers which has received some recent attention. Studies of 'journeys to crime' have produced somewhat unexceptional findings and a distance-decay function, with many offences committed in the offender's own neighbourhood, seems general. This local nature of offending was confirmed by Baldwin and Bottoms (1976), who found little variation by social class or degree of recidivism, though older offenders tended to travel further. Haring (1972) summarised a series of American studies which showed some distance variations according to type of offence. For narcotics offences offenders had travelled an average of 2.17 miles, for petty larceny 1.83 miles, for burglary 0.77 miles and for vandalism 0.62 miles. Pyle *et al.* (1974) showed that whereas burglars travelled an average of 7.3 miles to high income areas, they only moved 0.93 miles to low-income areas; violent offenders had typically travelled only short distances. More recently, Phillips (1980) confirmed the general distance-decay and identified four distinct types of journey pattern for assault, vandalism, petty larceny and drug offences.

A number of authors have tried conceptually or empirically to identify the criminal's image of environment and so understand his responses to it. Scarr (1972) referred to an 'opportunity matrix' in which offenders perceive specific areas of the city on a kind of rating system; Letkemann (1973) developed the concept of 'casing' as a systematic assessment of target possibilities; Carter and Hill (1980) reported interviews with offenders and non-offenders in Oklahoma City from which considerations of 'familiarity' and 'strategy' emerged. For black offenders, crime patterns were predominantly influenced by familiarity evaluations, white offenders were equally influenced by familiarity and ease of committing offences. Studies which seek to understand criminal behaviour in this way have considerable potential and can be focused upon victims rather than offenders. Ley (1974) used interviews with residents in a Philadelphia neighbourhood to identify 'stress areas' which people tended to avoid as dangerous or unsafe, Smith and Patterson (1980) conducted a similar but more limited exercise in Norman, Oklahoma. The question of how residents of 'vulnerable' areas perceive their neighbourhoods and react to the possibility of victimisation has important connotations both for crime control and for the whole range of quality of life research.

Conclusions

The topics of crime and delinquency have only recently become recognised elements for research and teaching in urban geography and, in common with other social problem themes, they raise new kinds of research issues and multi-stranded possibilities for further study. From the long-established literature of criminology it has been possible to identify and advance a set of approaches which basically forms a spatial ecology perspective. Significant advances have been made towards setting this kind of research on a more systematic basis and towards improving the basic data storage and retrieval systems by which criminal statistics are recorded and maintained. At the complex interface with sociological perspectives on crime causation theory, a geography of crime which focuses upon area or neighbourhood effects has a useful but limited contribution to make. New lines of interest in 'managerial' roles may help understand why problem 'areas' emerge in the urban housing market and, with reference to a more direct interest in policing and the judiciary, Lowman (1982) urges geographers towards a much greater emphasis upon the social control system in their attempts to achieve a greater understanding of crime.

Several writers (Winchester, 1978) have sought to direct the interest of geographers towards the study of where offences occur rather than where offenders live. The advantages of this are clear both in terms of the possibilities of relating criminal behaviour to environmental variables and also of producing research results which may be of relevance to policy. Geographers are presented with many opportunities in the study of crime. There is scope to apply the various methodological perspectives of urban geography from the rigorous data analyses of spatial analysis to the humanistic insights offered by cognitive mapping and a probing of the values attached to space and place. Offenders, offences, victims, police and judiciary all form subject matter to which these perspectives may be applied. Objectives of this research will be various but the overall aim must surely be to relate its findings to the practical needs of society and to the reform of both the local environments in which crime and criminals are found and of the conditions which produce them.

References

Baldwin, J. (1975) 'Urban Criminality and the Problem Estate', *Local Government Studies*, *1*, 12–20

Baldwin, J. and Bottoms, A.E. (1976) *The Urban Criminal*, Tavistock, London

Boggs, S.L. (1966) 'Urban Crime Patterns', *American Sociological Review*, *30*, 899–908

Bradbury, J. (1982) 'An Investigation of the Spatial Distribution of Crime in Greater Nottingham', unpublished PhD thesis, University of Nottingham

Brantingham, P.J. and P.L. (1975) 'Residential Burglary and Urban Form', *Urban Studies*, *12*, 273–84

Brantingham, P.J. and P.L. (1980) 'Crime, Occupation and Economic Specialisation' in D.E. Georges-Abeyie and K.D. Harries (eds.) *Crime: A Spatial Perspective*, Columbia University Press, New York, pp. 93–108

Brown, M.J., McCulloch, J.W. and Hiscox, J. (1972) 'Criminal Offences in an Urban Area and the Associated Social Variables', *British Journal of Criminology*, *12*, 250–68

Caplow, T. (1949) 'The Social Ecology of Guatemala City', *Social Forces*, *28*, 113–35

Carr-Hill, R.A. and Sterns, N.H. (1979) *Crime, the Police and Criminal Statistics*, Academic Press, London

Carter, R.L. and Hill, K.Q. (1980) 'Area Images and Behaviour; An Alternative Perspective for Understanding Urban Crime', in Georges-Abeyie and Harries (eds.) *Crime: A Spatial Perspective*, pp. 193–204

Castle, I.M. and Gittus, E. (1957) 'The Distribution of Social Defects in Liverpool', *Sociological Review*, 5, 43–64

Cloward, R.A. and Ohlin, L.E. (1960) *Delinquency and Opportunity*, Free Press, Chicago

Corsi, T.M. and Harvey, M.E. (1975) 'The Socio-economic Determinants of Crime in the City of Cleveland', *Tijdschrift voor Economische Sociale Geografie*, 66, 323–36

Damer, S. (1974) 'Wine Alley: the Sociology of a Dreadful Enclosure', *Sociological Review*, 22, 221–48

Davidson, R.N. (1981) *Crime and Environment*, Croom Helm, London

Edwards, A. (1973) 'Sex and Area Variations in Delinquency in an English City', *British Journal of Criminology*, 13, 121–37

Evans, D.J. (1980) *Geographical Perspectives on Juvenile Delinquency*, Gower, Farnborough

Fletcher, J. (1849) 'Moral Statistics of England and Wales', *Journal of the Royal Statistical Society of London*, 12, 151–81, 189–335

Forman, R.E. (1963) 'Delinquency Rates and Opportunities for Subcultural Transmission', *Journal of Criminal Law, Criminology and Police Science*, 54, 317–21

Gill, O. (1977) *Luke Street*, Macmillan, London

Guerry, A.M. (1833) *Essai sur la Statistique Morale de la France avec Cartes*, Crochard, Paris

Hamnett, C. (1979) 'Area-based Explanations: A Critical Appraisal' in D.T. Herbert and D.M. Smith (eds.) *Social Problems and the City: A Geographical Perspective*, Oxford University Press, London, pp. 244–60

Haring, L.L. (1972) (ed.) *Summary Report of Spatial Studies of Juvenile Delinquency in Phoenix, Arizona*, Geography Dept., Arizona State University

Harries, K.D. (1974) *The Geography of Crime and Justice*, McGraw-Hill, New York

Harries, K.D. (1980) *Crime and the Environment*, Charles C. Thomas, Springfield, Illinois

Harries, K.D. and Brunn, S.D. (1978) *The Geography of Laws and Justice*, Praeger, New York

Hayner, N.S. (1946) 'Crimogenic Zones in Mexico City', *American Sociological Review*, 11, 428–38

Haynes, R.M. (1973) 'Crime Rates and City-size in America', *Area*, 5, 162–5

Herbert, D.T. (1976) 'The Study of Delinquency Areas: a Social Geographical Approach', *Transactions of the Institute of British Geographers*, new series, 1, 472–92

Herbert, D.T. (1982) *The Geography of Urban Crime*, Longman, London

Hindess, B. (1973) *The Use of Official Statistics in Sociology*, Macmillan, London

Hirschi, T. and Selvin, H.C. (1967) *Delinquency Research: An Appraisal of Analytic Methods*, Macmillan, London

Jones, C.D. (1934) *Social Survey of Merseyside*, Liverpool University Press

Kitsuse, J.I. and Cicourel, A.K. (1963) 'A Note on the Use of Official Statistics', *Social Problems*, 11, 131–9

Kobrin, S. (1951) 'The Conflict of Values in Delinquency Areas', *American Sociological Review*, 16, 653–61

Lander, B. (1954) *Towards and Understanding of Juvenile Delinquency*, Columbia University Press, New York

Letkemann, P. (1973) *Crime as Work*, Prentice-Hall, New Jersey

Ley, D. (1974) *The Black Inner City as Frontier Outpost*, Association of

American Geographers, Washington

Ley, D. and Cybriwsky, R. (1974) 'The Spatial Ecology of Stripped Cars', *Environment and Behaviour*, *6*, 53–67

Lottier, S. (1938) 'Distribution of Criminal Offences in Sectional Regions', *Journal of Criminal Law, Criminology and Police Science*, *29*, 329–44

Lowman, J. (1982) 'Crime, Criminal Justice Policy and the Urban Environment' in D.T. Herbert and R.J. Johnston (eds.) *Geography and the Urban Environment*, vol. 5, Wiley, London

Mack, J. (1964) 'Full-time Miscreants, Delinquent Neighbourhoods and Criminal Networks', *British Journal of Sociology*, *15*, 38–53

Mannheim, H. (1965) *Comparative Criminology*, Routledge and Kegan Paul, London

Matza, D. (1964) *Delinquency and Drift*, Wiley, New York

Mayhew, H. (1862) *London Labour and the London Poor*, Griffin-Bohn, London

Mayhew, P., Clarke, R.V.G., Sturman, A. and Hough, J.M. (1976) *Crime as Opportunity*, Home Office Research Unit, Study no. 34, HMSO, London

Mays, J.B. (1963) 'Delinquency Areas: A Re-assessment', *British Journal of Criminology*, *3*, 216–30

Miller, W.B. (1958) 'Lower-class Culture as a Generating Milieu of Gang Delinquency', *Journal of Social Issues*, *14*, 5–19

Morris, T.P. (1957) *The Criminal Area: A Study in Social Ecology*, Routledge and Kegan Paul, London

Newman, O. (1972) *Defensible Space*, Macmillan, New York

Phillips, P.D. (1972) 'A Prologue to the Geography of Crime', *Proceedings of the Association of American Geographers*, *4*, 59–64

Phillips, P.D. (1980) 'Characteristics of Typology of the Journey to Crime' in Georges-Abeyie and Harries (eds.) *Crime: A Spatial Perspective*, pp. 167–80

Phillipson, M. (1971) *Sociological Aspects of Crime and Delinquency*, Routledge and Kegan Paul, London

Pyle, G.F. (1974) (ed.) *The Spatial Dynamics of Crime*, Geography Research Paper 159, University of Chicago

Rengert, G.F. (1980) 'Spatial Aspects of Criminal Behaviour' in Georges-Abeyie and Harries (eds.) *Crime: A Spatial Perspective*, pp. 47–57

Repetto, T.A. (1974) *Residential Crime*, Ballinger, Cambridge, Mass.

Rock, P. (1973) *Deviant Behaviour*, Hutchinsons University Library, London

Scarr, H.A. (1972) *Patterns of Burglary*, US Dept. of Justice, Washington

Schmid, C.F. (1960) 'Urban Crime Areas', *American Sociological Review*, *25*, 527–54 and 655–78

Scott, P. (1972) 'The Spatial Analysis of Crime and Delinquency', *Australia Geographical Studies*, *10*, 1–18

Shaw, C.R. (1930) *The Jack-roller*, University of Chicago Press, Chicago

Shaw, C.R. and McKay, H.D. (1942) *Juvenile Delinquency and Urban Areas*, University of Chicago Press, Chicago, revised edn., 1969

Smith, C.J. and Patterson, G.E. (1980) 'Cognitive Mapping and the Subjective Geography of Crime' in Georges-Abeyie and Harries (eds.) *Crime: A Spatial Perspective*, pp. 205–18

Smith, S.J. (1981) 'Negative Reaction: Crime in the Inner City' in C. Peach, S.J. Smith and V. Robinson (eds.) *Ethnic Segregation in Cities*, Croom Helm, London, pp. 35–58

Stedman-Jones, G. (1971) *Outcast London: A Study in the Relationships between Classes in Victorian Society*, Oxford University Press

Sutherland, E.H. (1940) 'White Collar Criminality', *American Sociological Review*, *5*, 1–12

Sutherland, E.H. and Cressey, D.R. (1970) *Principles of Criminology*, Lippincott, Philadelphia

Taylor, L. (1973) 'The meaning of environment', Ch. 2, pp. 54–63 in C. Ward (ed.) *Vandalism*, Architectural Press, London

Timms, D.W.G. (1965) 'The Spatial Distribution of Social Deviants in Luton, England', *Australia and New Zealand Journal of Sociology*, *1*, 38–52

Tobias, J.J. (1967) *Crime and Industrial Society in the Nineteenth Century*, Penguin, Harmondsworth

Waller, I. and Okihiro, N. (1978) *Burglary: the Victim and the Public*, University of Toronto Press, Toronto

Wilson, H. (1980) 'Parental Supervision: A Neglected Aspect of Delinquency', *British Journal of Criminology*, *20*, 203–35

Winchester, S.W. (1978) 'Two Suggestions for Developing the Geographical Study of Crime', *Area*, *10*, 116–20

4 ETHNICITY

G.C.K. Peach

In 1980 the magnificent *Harvard Encyclopedia of American Ethnic Groups* was published (Thernstrom *et al.*, 1980). It is a large book published in quarto format and heavy, weighing 2.5 kilograms. It is also very long, running to 1,076 double-columned pages. Yet, on the first page of the book the editors confess: 'there is as yet no consensus about the precise meaning of ethnicity'. Ethnicity is an increasingly complex phenomenon. All the groups treated in the *Harvard Encyclopedia* were characterised by some of the following features (given in this order) although in combinations that vary considerably.

1. common geographic origin;
2. migratory status;
3. race;
4. language or dialect;
5. religious faith or faiths;
6. ties that transcend kinship, neighbourhood and community boundaries;
7. shared traditions, values and symbols;
8. literature, folklore and music;
9. food preferences;
10. settlement and employment patterns;
11. special interests in regard to politics in the homeland and in the country of settlement;
12. institutions that specifically serve and maintain the group;
13. an internal sense of distinctiveness;
14. an external perception of distinctiveness.

The degree to which these features characterise any group varies considerably with the size and specific history of the group. Ethnic groups persist over long periods, but they also change, merge or dissolve. New groups come about through the process known as ethnogenesis; others disappear.

In explaining ethnic groups however it is vital to understand that ethnicity is the linkage of two separate structures, the biological and

103

the cultural, into a fused entity. By the biological element I mean predominantly the racial characteristics of man. There is, in fact, rather little genetic variation in mankind despite the large range of different phenotypes (Negroid, Caucasoid etc.). Physical anthropologists calculate that if the world population were eliminated apart from a small group of people in an isolated area such as upland New Guinea, that population would contain the whole range of mankind's genetic pool. Despite this relative genetic homogeneity, distinct phenotypes exist and these phenotypes perpetuate themselves genetically through procreation by matched couples.

Cultural aspects of group identity, unlike the genetic characteristics, are transmitted environmentally. It is surprising the extent to which it is believed that cultural attributes, such as language, are genetically rather than environmentally transmitted. There is a rather neat, if apocryphal story of the Emperor Charles V of Austria who formulated a scientific experiment to establish whether Greek or Hebrew was the original language of mankind. Two babies were taken from their mothers at birth and given to nurses who were instructed not to speak a single word to their charges. The object of the experiment was that when the children began to speak, if they spoke Hebrew or if they spoke Greek they would produce irrefutable evidence for the original language. Regrettably, the children lacking maternal interaction died, ruining the experiment. The falsity of the design assumptions of the experiment is obvious: children learn to speak by being spoken to. Even Chomsky's belief in the 'deep structure' of language, that allows children to learn this immensely complicated process, does not prescribe what the learnt language is. Language in people is learned not innate, unlike birdsong in birds (which can sing even if hatched and reared in total isolation). Language, accent and the other elements of ethnic culture are transmitted from the group to individuals born into the group.

Ethnicity I view as the linkage of a particular cultural mode with a particular genetic stock: the genetic stock and the cultural units are more generally distributed but each combination is unique. Thus Caucasoid whites form the basic genetic element to which a whole series of European cultural traits attach themselves. They thus form a whole series of distinct European ethnicities. On the other hand, a culture such as Spanish may be attached to a whole series of different racial elements from Latin America to the Philippines to form a series of distinct ethnicities. In the first case race is held constant and culture varies; in the second culture is held constant and race varies.

But in both cases the result is to produce distinct ethnicities.

It could be argued that culture has a half life and is much more subject to decay and manipulation while race is more constant. Race will persist irrespective of the cultural superstructure. But even so, cultures are powerful streams that constantly replete themselves. The result, however, of the genetic reproduction of race and the environmental reproduction of culture is that phenotypes have been invested with strong cultural significance.

If ethnicity is difficult to define then assimilation, the process by which it is supposed to be modified is even more problematical. In order to cut through the Gordian knot of definition, I follow Lieberson's (1963) definition that assimilation has taken place when it is not possible to predict more about an individual or a group through knowing his or their ethnic origin than it is for any member of the population as a whole.

From a geographical point of view, writing on assimilation can be divided into two groups, the spatial and the aspatial. The spatial school argued that degrees of ethnic clustering were fundamental to understanding the degree of assimilation of groups into the societies into which they had migrated. The aspatial school, on the other hand viewed segregation or dispersal as casual outcomes rather than causal factors. Segregation was simply another aspect of a set of attitudes of one group to another which might have some mechanical impact on the working of society, but which was not basic to analysis. The role of spatial segregation in ethnicity is therefore a key divisive factor in the literature. Ethnic segregation, the geographical differentiation of society in urban areas particularly has become a central focus of contemporary social geography.

The Aspatial School of Ethnic Assimilation

Perhaps the most influential proponent of the aspatial school is Milton Gordon whose book *Assimilation in American Life* was published in 1964. Gordon argued that American ideas on assimilation had passed through four major phases: Anglo-Conformism, the Melting Pot, the Triple Melting Pot and Structural Pluralism.

Anglo-Conformism

In Anglo-Conformism, which was dominant as a political philosophy during the nineteenth century, it was held that the characteristics of

American society were white, Anglo-Saxon and Protestant (WASP). These social characteristics of the core population were regarded as immutable and immigrant ethnic groups were expected to discard those elements of their ethnicity which differed and adopt those of the core society. The core society was rigid; immigrant groups were to be flexible and adaptive.

The Melting Pot

In the early part of the twentieth century the rival view of the melting pot gained in popularity. This aimed at flexibility and convergence on both the part of the core society and the immigrant ethnic groups. In practice, the change appeared more substantial than it was. The melting pot was still effectively a rather white society. Its 'racial' boundaries turned out to be cultural rather than genetic: it was to extend to the eastern and southern Europeans rather than the blacks. If we take Anglo-Saxonism in WASP (White Anglo-Saxon Protestant) to stand for the English language and democratic institutions, then the effect of melting was to produce an enriched English (as spoken by Henry Kissinger, perhaps) rather than Esperanto. Similarly, the effect of the melting pot on the Protestantism of the WASP culture was not to produce a syncretist new religion of Judaeo-Christianity. Protestants, Catholics and Jews were poured into the melting pot and Protestants, Catholics and Jews emerged.

The Triple Melting Pot

It could be argued, therefore that the effect of the melting pot was to produce an acceptable religious pluralism rather than true fusion. However, Kennedy produced an analysis of ethnic intermarriage in New Haven, Connecticut (Kennedy, 1944; and 1952) which claimed that national ethnic differences were disappearing but within existing religious bulwarks. Kennedy claimed that Protestant nationalities such as the British, the Germans and the Scandinavians were coalescing to form a Protestant melting pot; the Irish, the Italians and the Poles were similarly merging to form a distinct Catholic melting pot while eastern and western European Jews were merging to form a Jewish melting pot.

Structural Pluralism

Kennedy's triple melting pot left no room for the blacks. Black Americans were predominantly Protestant, yet as Glazer and Moynihan (1963) indicated, they did not form part of a Protestant pot. Glazer and Moynihan also indicated the ambivalent position of the Puerto

Ricans. Puerto Ricans were Roman Catholic but racially mixed showing a range of phenotypes from Negro to Caucasian. Perhaps 5 per cent of the Puerto Rican population was black and with the vast majority of the remaining population brown but showing varying degrees of negroid features in hair, nose and lip shape. Puerto Ricans were ethnically Puerto Ricans, but translated to New York where racial lines were drawn differently from Puerto Rico, Glazer and Moynihan speculated that four possible assimilation or accommodation strategies faced them: (1) they could become part of the Catholic melting pot; (2) they could become part of black America; (3) they might be torn apart with black Puerto Ricans becoming part of black America and white Puerto Ricans becoming part of Catholic America, or (4) they might remain as Puerto Ricans. It was this fourth possibility that led Glazer and Moynihan to examine whether groups which had been presumed to have been melted had, in fact, disappeared. Apart from the Germans, most other European ethnicities appeared to have substantial vitality, if rather modified from its first generational form. It appeared therefore that structural pluralism was a more accurate description of the American ethnic pattern than the melting pot.

Spatial Analysis of Ethnicity

Given the nature of the growth of the US population during the nineteenth and early twentieth centuries with substantial chain migration as a major element of growth (Thomas, 1954), ethnic areas became a common feature of American towns and cities (Ward, 1971; Philpott, 1978). The early work suggested that these ethnic areas would break up with time and diffuse. Work by Cressey (1938), Ford (1950) and Kiang (1968) did demonstrate the outward movement of the centres of gravity of ethnic groups in Chicago over a series of years.

The most profound contribution to the analysis of *why* spatial distribution should be significant to the process of assimilation at all, however, was given by Robert E. Park (1926) in a paper entitled 'The Urban Community as a Spatial Pattern and Moral Order'. In this Park argued that social and physical distances could be equated:

It is because geography, occupation, and all the other factors which determine the distribution of population determine so irresistibly and fatally the place, the group, and the associates with whom each of us is bound to live that spatial relations come to have, for the study of society and human nature, the importance which they do.

It is because social relations are so frequently and so inevitably correlated with spatial relations; because physical distances, so frequently are, or seem to be, the indexes of social distances, that statistics have any significance whatever for sociology. And this is true, finally, because it is only as social and physical facts can be reduced to, or correlated with, spatial facts that they can be measured at all. (Park, 1926; 1975, 30–31)

Although Park argued that measurement of the physical distance of one group from another should prove an important measure of the degree of acceptance of those groups for each other, he did not indicate the mechanism through which distributions might affect identity. The assumption was that the more a group became dispersed the more assimilated it became.

Pattern and Interaction

The force which underlay this intuitive assumption was the relationship of social interaction to the spatial patterning of participants. This is a rather high flown way of arguing the case for gravity decay. People tend to interact more with those who are close than with those who are further away. But as well as the mechanistic arguments of distance decay there are social preferences in interaction also. People interact more with those who are like them than those who differ from them. Segregation produces a situation in which those who are closest are also the most like. Thus interaction between those who are closest and most like reinforce each other (Peach, 1974a).

The clearest demonstration of this argument was made by Ramsøy (1966). She measured likeness in terms of social class and interaction in terms of marriage. She was able to demonstrate through the use of contingency tables that (1) brides and grooms overselected spouses of the same or adjacent occupational class and underselected those of more distant classes; (2) brides and grooms dramatically overselected (in relation to the statistical chances) spouses living within a mile or so of their residences and (3) they dramatically overselected those of the same or similar class living close.

The true importance of segregation for ethnic identity can be seen if the relationship between assimilation and dispersal is reversed and we look instead at the strategies for preserving ethnic identity. We have

argued that the cultural element of ethnicity is transmitted environmentally. If one wishes to preserve a mother tongue in a foreign country,
that mother tongue must be the natural language for everyday
activities. It must be used and spoken without self-consciousness with
all whom are met. It is only in a situation where all speak the same
language that the language is not noticed. The most important elements
in cultural identity are those which are taken for granted. Their
importance is revealed in the outside world when they are no longer
taken for granted. It is not simply language which is absorbed in this
way but all of the other social assumptions and values of the group:
religious attitudes and beliefs, attitudes to kin and outsiders.

It is becoming clear through the work of sociolinguists that
proximity and interaction not only inculcates individuals with particular
accents and vocabularies, but is part of the mechanism of linguistic
change and dialect development. Labov has shown how the key
members of a black Harlem gang not only lived close to one another
and had reciprocal high rankings of one another but had significant
vocabulary and syntactical differences from socially or geographically
peripheral members (Labov, 1972, 274–81).

Sutcliffe elaborates this point in a British context and shows how
young blacks have developed Jamaican Creole as a badge of ethnic
identity — as a private language to shut out the oppressive elements of
the white world. (Sutcliffe, 1982, 147–54).

Segregation has other utilities for the preservation of ethnic identity.
Ethnic concentration allows the traversing of the critical thresholds for
ethnic shops, ethnic churches or other religious institutions, clubs and
even schools. Thus as well as the intertwining of the informal friendship
ties and value reinforcing benefits of supportive members of the group,
there are also the more formal and institutional buttresses.

The Impact of the Index of Dissimilarity

Park's injunction about the importance of measurement, and of the
correlation of physical and social distance was taken up by a series of
writers who made a major contribution to an understanding of the
spatial basis of ethnicity. These writers include Duncan and Duncan
(1957), Lieberson (1963), Taeuber and Taeuber (1965) and Kantrowitz
(1973). The methodology which gave their analysis impetus was the use
of the index of dissimilarity (ID).

The key factor in testing Park's assertion about the relationship of physical and social distance was finding ways of measuring social distance and physical distance. Social distance presented relatively few problems. It could be measured attitudinally or behaviourally. In attitudinal tests, subjects were questioned on their evaluation of members of different groups: 'Would you live next door to an X?' 'Would you marry a Y?' 'Would you let your daughter marry a Y?' and so forth. Some investigators assessed attitudes by presenting their subjects with a physical scale, such as a ruler and asking them to demonstrate how close they felt on such a scale to group A or B or C (Timms, 1969). Behavioural measurement took the same kind of ideas as the attitudinal tests, but attempted to observe interactions instead of attitudes to hypothetical questions. Behavioural studies tried to assess the number of intermarriages between members of different groups or to measure the proportion of members of an immigrant group who learned the language of the country in which they had settled or to measure the percentage of the immigrant group who had become citizens of the new country and so on.

The measurement of physical distance, on the other hand, proved a very much more difficult task. Initially distance was conceived of as physical distance from the centre of a city as in the studies of Cressey and Ford already mentioned. The difficulty with such measurements was that they assumed that outward movement meant dispersal. Whereas this was empirically true of most groups it was not necessarily true; and it was empirically untrue of the Jews and the blacks. The Jewish population of US cities decentralised substantially without dispersing, that is to say, they relocated, but without breaking up. The black population showed an outward movement from the centre but largely by virtue of the physical extension of the ghetto. Thus its outward movement was not even marked by substantial relocation of the area of concentration.

It was apparent that physical distance should be conceived of in terms of the spatial mix of residential population. Putting this another way, physical distance should be conceived of in terms of the degree of residential segregation. This still begged many questions because concepts of segregation — although sharing the same vocabulary — were often radically different. For example, if we take Tiger Bay in Cardiff which contained almost all of Cardiff's pre-war black population (Little, 1947) one could argue two totally contradictory views about their degree of segregation. Little, for example, argued that the black population was highly segregated since it was confined to this single area. Other observers argued the opposite, since Tiger Bay contained a large white population. Thus since the black population of Tiger Bay lived with whites in a mixed area they were, according to this view, unsegregated while the majority of whites, who lived outside Tiger Bay with no black neighbours were truly segregated. Segregation, in other words, was a slippery concept and even when observers agreed on the form of the distribution, they disagreed fundamentally on what the measurement of that distribution should be (Peach, 1981).

Because the concepts of segregation were so different, it was inevitable that mathematical formulae, which reflected these different concepts, would also produce conflicting results. From the mid 1940s to the mid 1950s there was in the American sociological literature an index war which seemed to have been resolved by a review article by Duncan and Duncan (1955) in which they argued for the superiority of the index of dissimilarity. ID was similar conceptually to many of the indexes used by economists to measure inequality of income distributions in populations and the same index is used extensively by David Smith in *Human Geography: A Welfare Approach* in order to demonstrate unequal access to resources. The index was useful in having a verbal translation: the index figure which ranged from 0 to 100 could be translated as the percentage of group A which would have to shift from its area of residence in order to replicate perfectly the distribution of group B with which it was being compared (or *vice versa*).

This simple technique of measurement opened the gates of research to a massive advance in the understanding of ethnicity and assimilation. The most significant advance came in a paper by Duncan and Lieberson (1959) on ethnic segregation and assimilation. In this paper Duncan and Lieberson used residential segregation as the independent variable and correlated with it a series of behavioural variables which sought to measure assimilation over a range of eleven ethnic groups in Chicago. The behavioural variables which they employed included measures of centralisation, legal naturalisation, percentage of foreign-born whites able to speak English, median monthly rental, values of houses, the median number of school years completed by each group, median income of each group, the degree of occupational similarity with the native-born white population and a surrogate measure for the degree of intermarriage between the foreign-born and the native born. The results of these correlations all pointed with different degrees of strength towards the hypothesised direction. The more highly segregated, the lower the percentage able to speak English ($r = -0.83$); the more segregated, the less the inter-marriage ($r = -0.88$); the more segregated, the lower the median monthly rental ($r = -0.74$); the more segregated,

Table 4.1: Indices of Dissimilarity of Selected Foreign-born White Groups, Native White and Black Population, Chicago[a] 1930 and 1950

(Above diagonal, 1950 figures — below 1930: numbers at head of columns refer to countries numbered on stub)

Country of Origin	1	2	3	4	5	6	7	8	9	10	11	12
1. England and Wales	—	28.5	29.7	29.5	58.4	55.9	26.3	38.3	56.8	45.7	77.8	18.9
2. Eire	24.7	—	40.2	43.9	66.7	63.2	38.3	54.2	59.8	52.0	81.4	31.8
3. Sweden	30.7	42.4	—	32.3	67.8	66.0	38.8	54.0	66.2	60.9	85.5	33.2
4. Germany	35.0	44.1	35.1	—	55.9	47.2	21.3	47.2	54.3	54.3	85.4	27.2
5. Poland	64.6	68.3	73.5	57.7	—	43.5	47.3	58.3	50.8	52.6	90.8	45.2
6. Czechoslovakia	60.0	63.7	68.8	58.6	47.2	—	47.8	61.4	50.5	55.9	89.2	48.8
7. Austria	34.3	43.9	45.3	22.3	49.4	49.1	—	45.6	53.8	45.6	82.5	18.1
8. USSR	50.1	59.5	65.5	56.4	56.7	62.5	48.2	—	67.5	57.5	87.1	44.0
9. Lithuania	62.5	62.2	72.8	65.5	51.2	49.9	57.1	68.6	—	61.6	84.7	51.5
10. Italy	53.4	56.6	66.9	57.4	58.8	63.6	52.2	56.7	66.4	—	69.6	40.5
11. Negro	83.6	84.3	90.1	88.6	93.2	92.7	88.4	89.8	90.9	79.2	—	N.A.
12. Native Whites	19.1	31.8	34.0	26.0	50.8	51.9	25.0	49.8	57.0	48.3	N.A.	—

Note: a. For the 75 community areas.
Source: Compiled from Tables 1, 2 and 3, pp. 366–8, Duncan and Lieberson, 1959.

the lower the median number of school years (r = −0.65); the more segregated, the greater the difference in occupational structure between the immigrants and native whites (r = +0.62) (Duncan and Duncan, 1959; Peach, 1975a, p. 105).

Further work by Lieberson (1963) demonstrated that the patterns of ethnic segregation which he and Duncan had established for Chicago were characteristic of the ten other northern cities which he investigated. These results can be portrayed in tables of indexes of dissimilarity which measure the degree of residential overlap between any pair of ethnicities in the array. The tables may be read like mileage charts of distance between an origin and destination. For example, in Table 4.1 it can be seen that persons born in England and Wales were separated from those born in Eire by 28.5 points while they were separated from the Negro population by 77.8; Swedes were separated from the Italians by 60.9 and so on. Lieberson showed that, generally speaking, 'older' European groups from northern and western Europe had low degrees of segregation from each other and from the native born population while the 'newer' Europeans from southern and eastern Europe were more segregated from the native born population, the older European groups and each other. The black (Negro) population was most segregated of all, and from everyone.

Conflict between the Spatial and Aspatial Schools

While the aspatial school of assimilation studies had four successive models (from Anglo-Conformity through to Structural Pluralism), the spatial school had only one. As we have seen, the spatial school hypothesised that the more segregated the group, the less the assimilation; the more dispersed the group, the more the assimilation. Work by writers of the spatial school generally fitted one of the aspatial models. The work of Cressey, Ford, Kiang, Duncan and Lieberson generally conformed to the Melting Pot hypothesis, demonstrating general reduction in the degrees of segregation for most groups over time. Later writers, particularly Kantrowitz using the same model and rather similar results, interpreted them differently and claimed stability of ethnic segregation rather than dissolution. His results fitted into the Structural Pluralist model.

One consequence of Duncan and Lieberson's (1959) paper, however, was that an implicit conflict appeared between the spatial model of ethnic assimilation and the Triple Melting Pot hypothesis of Kennedy

(1944; 1952). The spatial model argued that interaction (of which intermarriage was an example) would be greatest between groups which were residentially intermixed and least between groups which were residentially segregated. Kennedy's Catholic melting pot in New Haven, Connecticut, proposed high intermarriage between the Irish, Poles and Italians. However, these groups appeared from the spatial analysis of other cities such as Chicago (see Table 4.1) to be relatively segregated from one another. The Irish had an ID of 66.7 with the Poles and 52.0 with the Italians; Italians and Poles were separated by an ID of 52.6. Although there were problems of inferring Polish ethnicity from Polish birthplace (many of the Polish-born in Chicago were Jewish), it seemed unlikely from these figures, which were similar to those for other cities where such analysis had been performed, that the Irish, Poles and Italians would constitute an intermarrying group.

Peach (1980) re-examined the New Haven data used by Kennedy (in which it was possible to define ethnicity rather than birthplace) and showed that the Triple Melting Pot did not exist. It was demonstrated also that there was a high positive correlation between the patterns of ethnic residential segregation and ethnic intermarriage. Ethnic groups which were residentially intermixed manifested high degrees of intermarriage (British Americans, Irish, Germans and Scandinavians) while groups which were relatively or highly segregated such as the Italians, the Jews and the black population showed high degrees of marriage into their own groups and relatively little out-marriage.

Ethnicity and Class

Marxist Interpretation of Ethnicity

Ethnicity presents difficult problems for Marxist analysis because while the phenomenon has enormous political potency, it is regarded as a diversionary and non-fundamental social categorisation. Ethnicity in the Marxist view is the product of historical determination in specific settings and as such has reality. However, the organisation of society to give prominence to this construct is regarded as counter-revolutionary. Ethnic differentiation outside a class analysis is regarded with some suspicion by Marxists because it acts as a divisive force within the working class preventing its solidarity in the face of capitalism. Studies of ethnicity are thus suspect also since they encourage the 'false consciousness'. Thus academic writers on ethnicity are stigmatised as counter-revolutionaries (Harvey, 1972, 1-12).

Ethnicity is profoundly disruptive not only of class unity but of national unity also. Northern Ireland represents one of the clearest cases of economic class interests being cross-cut by ethnic divisions. Northern Ireland regularly produced unemployment rates well in excess of the United Kingdom average. In mainland Britain in the past, its economic structure of dependence on declining traditional industries such as ship-building and textiles, would have produced a substantial Labour Party political representation on the model of other traditional industrial regions. In Northern Ireland, however, the Labour Party was never constituted and political parties represented competing nationalistic ideologies rather than Marxist political divides.

Emrys Jones in a classic work *A Social Geography of Belfast* (1960) showed how the population was segregated on class lines, but how the Roman Catholic/Protestant divide cross-cut the economic class divides. Segregation was sharpest in working class areas. Jones also demonstrated how the political climate between the communities affected the degree of segregation. In periods of relative calm, the degree of separation of the two groups relaxed but after periods of tension such as the bitter rioting which recurred in the city between 1920 and 1923 there were dramatic increases in the degree of segregation (Jones, 1960, 194-5).

Boal (1969) demonstrated behavioural effects of this segregation by examining the shopping, visiting and bus travel habits of the populations on either side of one of the sharpest segregation interfaces on the Shankill/Falls divide. His investigation took place before the escalation of the struggle between the two groups began in 1969, but even at this time the divide acted like a social Himalaya. Visiting, shopping and bus-catching activities fell away from the divide like streams from a watershed with very few activities crossing from one side to another. With the increase in violence, the communities have become even more spatially polarised with Roman Catholics and Protestants abandoning their homes in mixed areas dominated by the other community and with the minority Roman Catholic population in particular withdrawing itself into tight defensive core areas. Inner city ghettos like the Ardoyne or Unity Flats stand out like beleaguered garrisons in hostile territory. Peace fences have been built to separate the larger Falls and Springfield Road Catholic concentrations from the adjacent Shankill Protestant camp.

Murray and Boal's (1979) analysis of door-step murders (one form of sectarian atrocity in which the victim was chosen as a token representative of one of the communities) showed how the distribution of victims tended to lie along the boundary edges of segregated areas where the killers could make a rapid penetration and escape back to the safety of their own area.

Ethnicity in Classical Urban Models

In the classical models of the social morphology of cities, ethnicity appears as 'noise'. In Burgess's model of residential areas of Chicago (Burgess, 1924; Johnston, 1971, 75) for example, class is represented by the bold concentric rings outward from the Loop through the low class to the high class on the periphery. Ethnicity is represented by irregular labellings such as 'Deutschland', 'Little Sicily', 'China Town' or 'Black Belt' which sit uncomfortably and unconformingly upon the sharp class pattern. Similarly, the models of Hoyt and of Harris and Ullman are conceived of in economic terms. In other words, the classical urban models which dominated the thinking of geographers were economic determinist models.

Yet although social class is linked to the economic structure, it is by no means fully congruent with it. Economic class is structured by employment and earnings: social class is structured by living and spending. Economic class is structured by the mode of production: social life by life-style and consumption.

Impact of Income on Segregation

Lieberson (1963) demonstrated that segregation was not simply a function of income or occupation. Using a technique known as indirect standardisation, he was able to show that although income limited the areas in which members of ethnic groups could afford to buy or rent houses, income explained rather little of the choice *within* the areas that were open to them. The technique worked in the following way. If the distribution of each income group for a city were known for each sub-area of the city and if the percentage that each ethnic group formed of each income group for the city as a whole were known, it would be possible to produce a statistical expectation of the number of each ethnic group that would be found in each sub-area of the city, if that distribution were determined by income alone. From this distribution it would be possible to calculate how segregated that group would be if income alone were to determine its distribution. Thus, if, for example, Italians formed 1 per cent of the population earning over $20,000 per year in the city and 5 per cent of those between $10,000 and $20,000 and 10 per cent of those earning less than $10,000 then it would be possible to predict that for an area where there were 100 persons in each of these income categories there should be 16 Italians (1 + 5 + 10) and 284 non-Italians, if income alone had determined their distribution. Applying the same technique to the remaining areas of the city would produce the hypothetical distribution of Italians and the remainder of the population. From these figures the ID could be calculated and it would show how segregated Italians would be if income alone were to determine their distribution.

Lieberson, using Cleveland data for 1930 showed that while income differences 'explained' a significant amount of the moderate levels of 'old' European, for the newer groups it explained less. For example, the British were segregated from the American born whites of American-born parents by a modest ID of 18.7 while on economic grounds they would have manifested an even lower degree of segregation of 5.8. Observed and expected figures for Italians against the same group were 64.6 and 13.9 while for the non-whites the figures were 85.6 and 24.0 respectively (Lieberson, 1963, 88-9). Thus, it was clear that factors other than income explained the bulk of ethnic segregation. Blacks might be poor but poverty did not explain the extent to which they were segregated. On this reading, economic factors did not explain even one third of the segregation of the most highly segregated groups.

The question of the economic contribution to segregation was taken up in an important paper by Taeuber and Taeuber (1964; 1975) which uses Lieberson's method though without citing his earlier work. The Taeubers demonstrated that for Chicago in 1960, economic factors explained even less of the black segregation than did Lieberson's estimate for Cleveland in 1930. The observed level of non-white segregation was 83 while the 'expected' level was 10 so that only 12 per cent of racial segregation was attributable to income differences.

The Taeubers analysis indicated one extremely important hypothesis: that the level of segregation experienced by blacks must be attributable to a high level of racial prejudice that exceeded anything experienced by other ethnic groups. The same paper by the Taeubers offered further evidence for this hypothesis. They demonstrated that Puerto Ricans, who were poorer and less well educated than the blacks, were nevertheless less segregated than the blacks. The implication seemed to be that because Puerto Ricans were less negroid than blacks they were more acceptable than whites and therefore better able to disperse than blacks.

Kantrowitz (1969a) took this argument one stage further in his analysis of Negroes and Puerto Ricans in New York in 1960. He demonstrated that not only did the Puerto Rican distributions fringe the black ghettos (Puerto Rican foothills to the Negro mountains) but that the blacker Puerto Ricans were concentrated on the inner edges of these distributions where they overlapped the black ghetto while the lighter Puerto Ricans overlapped the white areas. Jackson (1981) has investigated the areas of black/Puerto Rican overlap, particuarly in Harlem and has shown that Puerto Ricans constitute a buffer between blacks and whites rather than a binding group. Puerto Ricans want to distance themselves from the blacks. This is most true perhaps of the darkest Puerto Ricans, 'the blacker the skin, the louder the Spanish' is one phrase which describes the attempt to widen the perceptual gap.

Despite the doubts over Kantrowitz's claim for the binding-group nature of the Puerto Ricans, he made significant contributions to the interpretation of degrees of ethnic segregation. Firstly, he (Kantrowitz 1969b, 1975) argued strongly that European ethnics were maintaining their degrees of segregation over time rather than dispersing and disappearing. Secondly, he argued that the segregation of blacks should not be seen in black and white terms but as part of a pattern of other inter-ethnic segregation. Thus, it was not simply a case of blacks being segregated from whites but of blacks being segregated from Italians and from Poles and from the Irish and of white ethnic groups also being segregated from each other. Kantrowitz demonstrated that even very similar groups such as the Norwegians and Swedes who shared some neighbourhoods in New York nevertheless had IDs of 45.4 in 1960. Norwegian segregation from USSR ethnic stock (which is largely Jewish in New York) with an ID of 70.7 approached the level of Norwegian segregation from Negroes at 87.7. 'If we assume that residential segregation numbers reflect degrees of cultural acceptance, we think it a fair speculation that Norwegian segregation from Negroes differs in degree, but not in kind, from their separation from Slavic Jews.' (Kantrowitz, 1969b; 1975, 145).

Kantrowitz's general argument is persuasive although there are objections to his particular example. Norwegians constitute a very small number in the population of New York City and there is evidence that high IDs can be caused through random factors where very small numbers or very fine gridding of the areal mesh is present (Peach, 1981). Nevertheless, Kantrowitz shows that not all high degrees of segregation can be explained directly by negative factors of discrimination. It is in his work that the roots of the disputes in British Geography over choice and constraint can be found.

Choice and Constraint

Because of the high degree of inter-relationships between variables, it is difficult to cut into this argument. Essentially the argument is about whether ethnic groups maintain their spatial cohesion in ethnic villages or in ethnic ghettos because they positively desire the proximity of their own ethnic group or whether they do so for the negative reasons of the hostility of outside society which forces them back onto their own resources. Clearly elements of both positive and negative forces may be present in any given situation so that the problem is then to distinguish whether one force is dominant and the other recessive (Peach, 1968). In a situation in which friends wish to stick together, is this a positive and dominant force simply because they are friends, or are they friends because they need mutual support in the face of external hostility?

Kantrowitz's analysis of inter-ethnic segregation in New York argued that a substantial part of inter-European ethnic segregation was due to the desire to maintain a separate culture rather than through external compulsion. The Italians, for example, consistently show significant levels of segregation in US cities with levels from the mid 40s to the high 50s in New York in 1960. Yet there is no suggestion that Italians are discriminated against in US society. Black segregation is, however, different in degree and although it is clear that if discrimination against blacks were to disappear from American society black segregation would not disappear, it is equally certain that it would drop from its median values in the high 70s and low 80s (Taeuber and Taeuber, 1965) down to substantially lower levels (Berry, 1971, 1975).

Similarly, viewing levels of segregation of immigrant groups in British society, Peach (1975b) has demonstrated that the most highly segregated immigrant group in London in 1971 was the Cypriot-born population with an ID against the native born of 54 in comparison with an ID of 51 for West Indians and the native born (Table 4.2). There is no suggestion that discrimination accounts for this high segregation of the Cypriots but rather, an independent desire of the group to maintain contact and perhaps even a rather narrow diffusion route northwards from the original port of entry of the group in Soho along the 29 bus route up the Tottenham Court Road (Oakley, 1982). Greek immigrants in Australian cities manifest similar high degrees of segregation. For example, Greeks in Sydney in 1966 had an index of 53 at ward level (Peach, 1974b, 233).

Discrimination in the housing market and their low economic

position clearly affected the range of housing open to West Indians

Table 4.2: London: Indices of Residential Dissimilarity, by Ward, for Selected Birthplaces, 1971

	1	2	3	4	5	6	7
1. England and Wales	—						
2. Irish Republic	29.92	—					
3. Old Commonwealth	41.34	39.03	—				
4. Caribbean	50.92	39.70	63.64	—			
5. India	38.18	34.46	46.64	50.30	—		
6. Pakistan	48.96	42.83	54.82	51.85	35.20	—	
7. Malta	34.42	35.74	48.54	46.47	43.54	45.34	—
8. Cyprus	53.57	47.42	63.55	47.22	59.88	61.26	54.13

Note: To find the index of dissimilarity between, for example, the Irish (2) and the Pakistanis (6), read row 6 against column 2. This indicates an ID of 42.83.
Source: Special tabulations supplied by Office of Censuses and Surveys of 1971 Census; Peach (1975b).

in Britain. Until 1971 very few West Indians were in council housing where between a fifth and a third of all British families were housed. Even when West Indians gained access to council housing it is evident that in many cities they were offered less good housing than equivalently placed white families. Parker and Dugmore's analysis of lettings in Greater London Council housing shows that whatever their category of rehousing, West Indians got on average one quality of housing lower than equivalently placed whites (GLC, 1976, 4; Parker and Dugmore, 1977/8; Peach and Shah, 1980).

Yet even so, there were quite marked levels of segregation within the West Indian population between West Indians from different islands (see Table 4.3). It can be seen that Jamaican segregation from Trinidadians approaches the level of West Indian segregation from the native population.

Perhaps the most interesting development of this area of research was carried out by Robinson (1979) on Asian settlement in Blackburn. The British census does not distinguish between the ethnic sub-groups from the sub-continent except insofar as they originate in India or Pakistan. The separation of Bangladesh from Pakistan occurred after the British 1971 census so that the Pakistan-born population hid the distinction between Bengalis and those born in West Pakistan. Asians who had migrated from East Africa were recorded as African born. Thus birthplace statistics were unhelpful in the analysis of intra South Asian ethnicity. Ethnicity among South Asians is multifarious. There is a total of over 1,500 languages on the sub-continent of which at least 15 are

Table 4.3: London: Indices of Dissimilarity of Selected Birthplaces, by Ward, 1971

	1	2	3	4	5	6	7
1. Barbados	—	27.83	26.41	31.73	31.31	17.45	49.11
2. Guyana		—	31.91	29.99	37.76	24.82	46.21
3. Jamaica			—	40.90	39.16	14.86	57.13
4. Trinidad and Tobago				—	37.59	31.62	44.45
5. Other Caribbean					—	26.34	54.36
6. Total Caribbean						—	50.92
7. England and Wales							—

Source: Special tabulations supplied by Office of Population Censuses and Surveys of 1971 Census.

Table 4.4: Indices of Dissimilarity of Religions in Blackburn, 1977 (at Ward Level)

	Muslims	Hindus	Sikhs	Non-Asian
Muslims	—	57.6	34.3	52.2
Hindus		—	39.7	60.7
Sikhs			—	61.8
Non-Asians				—

Source: Robinson (1979) Table 4.

as large as major European languages. Additionally there are the two major religions of Hinduism and Islam and the smaller, but still significant religions of Sikhism and Jainism. These variables cross cut one another and Hinduism is further sub-divided by caste. The antagonisms of these religious and linguistic groups on the sub-continent survive in a British context.

Robinson's achievement was to carry out a questionnaire survey of 1,693 out of the 2,098 Asian households on the electoral register in Blackburn and identify the ethnicity of each. From this he was able to calculate the degree of internal separation which had taken place between the various ethnic sub-groups from South Asia (see Table 4.4). From this, the degree of separation of Hindus from Muslims and Sikhs can be seen. He was also able to demonstrate that, where elements of ethnicity overlapped, some elements were more significant in producing residential overlap than others. For example, Punjabi-speaking Muslims were rather more sharply separated from Bengali-speaking Muslims with whom they had shared their erstwhile state of Pakistan than they were from Gujarati-speaking Muslims from India and East Africa.

The major proponent of the constraint school was Rex. In *Race, Community and Conflict* (Rex and Moore, 1967) Rex had argued that Asian immigrants were forced into a pariah section of the housing market by British racialism which accepted their labour but which denied them housing. Housing was available only in the twilight inner zones of cities where immigrants were forced to crowd and overcrowd since they were denied residence elsewhere. Rex was correct in much of his analysis. Overt discrimination against immigrants and their low position in the economic hierarchy made them victims of vicious exploitation in the private rental sector. Numerous studies testified to this exploitation (Milner Holland, 1965; Burney, 1967; PEP, 1967; Runnymede Trust, 1975; Smith and Whalley, 1976).

However, Dahya (1974) pointed out that not all of the ethnic clustering or even multi-occupation of Pakistanis should be interpreted as exploitation (Dahya, 1974). The 'myth of return' of many of the migrants made them less sensitive to living conditions in Britain since their object was to accumulate money to invest in their home villages. Under these circumstances, saving was of paramount importance. Renting from a member of one's own group was preferable to paying money out of the community; sharing meant the reduction of costs. Even if conditions were regarded as bad in absolute terms by the British, Dahya argued that in relative terms to the Pakistanis they might be better than they experienced in Pakistan.

The extreme constraint argument was developed by Aldrich, Cater, Jones, and McEvoy (1981) who argued that Asians in Britain were driven by the economic repression of British society to create an enclosed ethnic commercial market as the only way in which a minority of the group could seek economic advance. They examined Asian businesses in Bradford, Leicester and Ealing and showed the extent to which the businesses in the areas which they selected depended on Asian clienteles. Some of their conclusions are surprising. They found for instance that 'Indian' restaurants depended on Asian customers (Aldrich *et al.*, 1981). Mullins, working in Croydon (Mullins, 1979) produced opposite results. He found that Asian businesses were orientated towards the total rather than the ethnic market. He showed however, that Pakistani Muslims were most orientated towards the ethnic market and East African Hindus least. The predominance of Pakistani Muslims in Bradford might explain some of Aldrich *et al.*'s results in that city but there is a complete mismatch between what would be expected in Leicester and Ealing on the basis of Mullins' analysis and what is reported. Nearly half of Leicester's Asian population were East African Asians (Phillips, 1981, 104) whose educational attainments were generally reckoned to be high (Phillips, 1981, 108). Aldrich *et al.* report that East African Asians had low educational attainments. There thus appear to be doubts about the accuracy of some of the observations of the main proponents of the constraint school in the geographical literature.

Ethnicity as a Transactional Category

One of the conclusions that emerges from a study of ethnicity is that it is a transactional rather than categorical dimension. Ethnicity is not only the product of interaction within the group but it is not made explicit until transactions are undertaken with those outside the group. It is the ethnicity of others that makes us aware of our own. Because ethnicity is many layered and because the ethnicity of others is many layered, the degree to which ethnicity is displayed depends in part on the other actors. I may be Welsh in England, British in Germany and European in Thailand.

Some of the most interesting effects of the layering of ethnicity may be seen in the wake of decolonisation. Colonisation promotes the common front of the colonial peoples so that liberation struggles may promote the cooperation of groups, who are otherwise competitive, against a common enemy. The removal of colonial powers allows the next layer of ethnic differentiation to rise to the surface and to be eliminated and so on.

The former British Indian Empire gives a good example of this process. Hindus and Muslims cooperated to get rid of the British, but ethnicity forced the split of the country on religious lines in 1947. Pakistan, a two wing state with 1,000 miles of India separating the parts was held together by Islam and fear of India. It contained, however, at least two competing dominant ethnicities — the Punjabis and Bengalis who eventually split the state in 1971. India, rid of the British and the two dominant areas of Muslim concentration proceeded to split the existing provincial units and to amalgamate the princely states into new ethno-linguistic states, complete with border disputes. Further religious ethnicity forced the division of the Punjab State into new Sikh and non-Sikh states. Thus, as each pressure system was released, new lower denominators of ethnicity came into political play. Ethnicity, in other words, is a category which is in part a product of what it confronts.

Phenomenological Analyses of Ethnicity

Some of the most interesting investigations of ethnicity have examined the transactional nature of identity at the micro scale. Gerald Suttle's *Social order of the Slum* and David Ley's *Black Inner City as Frontier Outpost* are outstanding as examples of participant observation. For Ley, this approach was a counter-thrust to the number-crunching of the logical-positivists and a more human approach than that of the marxists who saw mankind in an economic mechanistic and determined world. Ley's study gave the feeling of place and a contextual understanding of life in the north Philadelphia ghetto. One of the incidents that Ley uses to illustrate the contextual role of ethnicity was when he was ignored in a road by a teenage gang who were challenging all other black youths who passed. Ley, the white, was not part of their life-world — he was a non-entity for their system and for them he did not exist.

Conclusion

Ethnicity is one of the most powerful dividers of society. Class differences are subsumed within it far more than it is subsumed within class. The classical urban models which added ethnicity as a social garnish on an economic class recipe may now have to have the balance of the ingredients reviewed.

Ethnicity teaches us, perhaps, that it is not the spatial organisation of society that is the focal point of geography but the social organisation of space.

There are three competing paradigms for the spatial investigation of ethnicity, the positivist, the marxist and the phenomenologist. The positivist paradigm has been the strongest and longest-lived in the field so far and has yielded through the use of the index of dissimilarity one of the most successful examples of cumulative social science. It treated ethnicity, however, as if it were a categorical category rather than a negotiated, transactional status. It is in this less well defined area that perhaps the new impetus for the study of urban ethnicity will come.

References

Aldrich, H.E., Cater, J.C., Jones, T.P. and McEvoy, D. (1981) 'Business development and self-segregation: Asian Enterprise in Three British Cities', 170–90 in Peach, C., Robinson, V. and Smith, S. *Ethnic Segregation in Cities*,

Croom Helm, London

Berry, B.J.L. (1971) 'Monitoring Trends, Forecasting Change and Evaluating Goal Achievements: The Ghetto v. Desegregation Issue in Chicago as a Case Study' in Chisholm, M., Frey, A.E. and Haggett, P. (eds.) *Regional Forecasting*, Colston Papers no. 22, Butterworth, London, reprinted in Peach, C. (ed.) *Urban Social Segregation*, Longman, London

Boal, F.W. (1969) 'Territoriality on the Shankhill Falls Divide, Belfast', *Irish Geography*, *6*, pp. 30–50

Burgess, E.W. (1924) 'The Growth of the City: An Introduction to a Research Project', Publications, *American Sociological Society*, *18*, 85–97

Burney, E. (1967) *Housing on Trial*, Oxford University Press, London

Cressey, P.F. (1938) 'Population Succession in Chicago: 1898–1930', *American Journal of Sociology*, *44*, 1–19

Dahya, B. (1974) 'The Nature of Pakistani Ethnicity in Industrial Cities in Britain' in Cohen, A. (ed.) *Urban Ethnicity*, Tavistock, London

Duncan, O.D. and Duncan, B. (1955) 'A Methodological Analysis of Segregation Indexes', *American Sociological Review*, *20*, 210–17

Duncan, O.D. and Duncan, B. (1957) *The Negro Population of Chicago*, Chicago University Press, Chicago

Duncan, O.D. and Lieberson, S. (1959) 'Ethnic segregation and assimilation', *American Journal of Sociology*, *64*, 364–74. Reprinted in Peach, C. (ed.) (1975) *Urban Social Segregation*, Longman, London

Ford, R.G. (1950) 'Population seccession in Chicago', *American Journal of Sociology*, *56*, 151–60

Glazer, N. and Moynihan, D.P. (1963) *Beyond the Melting Pot*, The MIT Press, Cambridge, Mass.

Gordon, M.M. (1964) *Assimilation in American Life*, Oxford University Press, New York

GLC (1976) *Colour and the Allocation of GLC Housing*: The report of the GLC lettings survey, 1974–75. Research Report 21, John Parker and Keith Dugmore, London, Greater London Council

Harris, C.D. and Ullman, E.L. (1945) 'The Nature of Cities', *Annals of the American Academy of Political and Social Science*, *242*, 7–17

Hoyt, H. (1939) *The Structure and Growth of Residential Neighbourhoods in American Cities*, Federal Housing Association, Washington

Jackson, P. and Smith, S.J. (eds.) (1981) *Social Interaction and Ethnic Segregation*, Institute of British Geographers Special Publication, no. 12, Academic Press, London

Johnston, R.J. (1971) *Urban Residential Patterns*, G. Bell & Sons Ltd, London

Jones, E. (1960) *A Social Geography of Belfast*, Oxford Univeristy Press, Oxford

Kantrowitz, N. (1969a) *Negro and Puerto Rican Populations of New York City in the Twentieth Century*, American Geographical Society, New York

Kantrowitz, N. (1969b) 'Ethnic and Racial Segregation in the New York Metropolis', *American Journal of Sociology*, *74*, 685–95. Reprinted in Peach, C. (ed.) (1975) *Urban Social Segregation*, Longman, London

Kantrowitz, N. (1973) *Ethnic Residential Segregation in the New York Metropolis: Residential Patterns Among White Ethnic Groups, Blacks and Puerto Ricans*, Praeger, New York

Kennedy, R.J.R. (1944) 'Single or Triple Melting Pot? Intermarriage Trends in New Haven, 1870–1940', *American Journal of Sociology*, *49*, 331–9

Kennedy, R.J.R. (1952) 'Single or Triple Melting Pot? Intermarriage in New Haven, 1870–1950', *American Journal of Sociology*, *58*, 56–9

Kiang, Y.-C. (1968) 'The distribution of ethnic groups in Chicago', *American Journal of Sociology*, *74*, 292–5

Labov, W. (1972) *Language in the Inner City; Studies in the Black English Vernacular*, Blackwell, Oxford

Ley, David (1974) *The Black Inner City as Frontier Outpost* Monograph 7, Assoc. Am. Geographers, Washington

Lieberson, S. (1964) *Ethnic Patterns in American Cities*, Free Press of Glencoe, New York

Little, K. (1947) *Negroes in Britain*, Kegan Paul, Trench, Trubner & Co., London

Milner Holland Committee (1965) *Committee on Housing in Greater London*, HMSO, London

Mullins, D. (1979) 'Asian Retailing in British Cities', *Working Note* no. 9, Graduate Geography Department, LSE, London

Murray, R. and Boal, F.W. (1979) 'The Social Ecology of Urban Violence' in Herbert, D.T. and Smith, D.M. (ed.) *Social Problems and the City*, 139-57, Oxford University Press, Oxford

Oakley, R. (1982) 'Changing patterns of distribution of Cypriot settlement', unpublished paper, Bedford College, London

Park, R.E. (1926) 'The Urban Community as a Spatial Pattern and a Moral Order' in E.W. Burgess (ed.) *The Urban Community*, reprinted in Peach, C. (ed.) (1975) *Urban Social Segregation*, Longman, London

Parker, J. and Dugmore, K. (1977/8) 'Race and the allocation of public housing – a GLC Survey', *New Community*, *6*, *1 & 2*, 27-40

Peach, C. (1968) *West Indian Migration to Britain: A Social Geography*, London, Oxford University Press for the Institute of Race Relations

Peach, C. (1974a) 'Homogamy, Propinquity and Segregation: A Re-evaluation', *American Sociological Review*, *39*, 636-41

Peach, G.C.K. (1974b) 'Ethnic Segregation in Sydney and Intermarriage Patterns', *Australian Geographical Studies*, *12*, 219-29

Peach, C. (ed.) (1975a) *Urban Social Segregation*, Longman, London

Peach, G.C.K. (1975b) 'Immigrants in the inner city', *Geographical Journal*, *141*, 3, 372-9

Peach, C. (1980) 'Ethnic Segregation and Intermarriage', *Annals of the Association of American Geographers*, *70*, 3, 371-81

Peach, C. and Shah, S. (1980) 'The Contribution of Council House Allocation to West Indian Desegregation in London, 1961-71', *Urban Studies*, *17*, 333-41

Peach, C. (1981) 'Conflicting Interpretations of Segregation' in Jackson, P. and Smith, S.J. (eds.) *Social Interaction and Ethnic Segregation*, 19-33. Special Publication no. 12, Institute of British Geographers, Academic Press, London

Phillips, D. (1981) 'The Social and Spatial Segregation of Asians in Leicester' in Jackson, P. and Smith, S.J. (eds.) *Social Interaction and Ethnic Segregation*, 101-21

Philpott, S. (1977) 'The Montserrations: Migration, Dependency and the Maintenance of Island Ties in England' in Watson, J. (ed.) *Between two Cultures: Migrants and Minorities in Britain*, Blackwell, Oxford

Philpott, T.L. (1978) *The Slum and the Ghetto: Neighborhood Deterioration and Middle Class Reform, Chicago, 1880-1930*, Oxford University Press, New York

PEP (Political and Economic Planning) (1967) *Racial Discrimination*, Political and Economic Planning, London

Ramsøy, N.R. (1966) 'Assortative Mating and the Structure of Cities', *Am. Soc. Review 31*, *6*, 773-86. Reprinted in Peach, C. (1975) *Urban Social Segregation*, Longman, London

Rex, J.A. and Moore, R. (1967) *Race, Community and Conflict*, Oxford University Press, Oxford

Robinson, V. (1979) *The Segregation of Asians within a British city: Theory and Practice*, Research Paper 22, Oxford, School of Geography

Robinson, V. (1980) 'Correlates of Asian Immigration: 1959–1974', *New Community*, *8*, *1 & 2*, 115–22

Runnymede Trust (1975) *Race and Council Housing in London*, Mimeo, London

Smith, D. and Whalley, A. (1976) *Racial Minorities in Public Housing*, Political and Economic Planning, London

Smith, D.M. (1977) *Human Geography: A Welfare Approach*, Arnold, London

Sutcliffe, D. (1982) *British Black English*, Blackwell, Oxford

Suttle, G. (1968) *The Social Order of the Slum: Ethnicity and Territory in the Inner City*, Chicago University Press, Chicago

Taeuber, K. and Taeuber, A. (1964) 'The Negro as an Immigrant Group', *American Journal of Sociology*, *69*, 374–82, reprinted in Peach, C. (ed.) *Urban Social Segregation*, Longman, London

Taeuber, K. and Taeuber, A. (1965) *Negroes in Cities: Residential Segregation and Neighborhood Change*, Aldine, Chicago

Thernstrom, S., Orlov, A. and Handlin, O. (1980) *Harvard Encyclopedia of American Ethnic Groups*, The Bellmays Press of the Harvard University Press, Cambridge, Mass.

Thomas, B. (1954) *Migration and Economic Growth*, Cambridge University Press, Cambridge

Timms, D.W.G. (1969) 'The Dissimilarity Between Overseas-Born and Australian-Born in Queensland: Dimensions of Assimilation', *Sociology and Social Research 53*, *3*, 363–74

Ward, D. (1971) *Cities and Immigrants*, Oxford University Press, New York

5 URBAN GOVERNMENT AND FINANCE

R.J. Johnston

Separate government for urban places has a long tradition in Western Europe, where it resulted from the conflict between two economic elites – the landed aristocracy and the merchant traders. Within Europe, as Hechter and Brustein (1980) make clear, the relative importance of the two varied according to the dominant mode of production. In two main zones – that dominated by feudalism, encompassing what is now the heartland of the EEC, and that dominated by petty commodity production, comprising much of Italy, Southern France, Catalonia, Switzerland, southern Germany and Schleswig-Holstein – towns were a fundamental component of economic life, and were the foundation for the development of the European state system (Johnston, 1982). In some cases, the towns formed separate city-states but in most, especially in the feudal zone, they were separate from but not independent of their hinterlands.

The creation of separate urban local governments reflected the desire by those operating the urban economies – the merchants and artisans – to establish monopolies, to limit entry to their professions, and to remove themselves from loyalty to the rural landowners (the lords of the manor). Separate status was obtained by the warrant of the over-lord or monarch; in Britain this involved the granting of a Charter which allowed the incorporation of a Municipal Borough and gave the burghers the right to control defined elements of the urban economy. In most cases, this included the operation of a market and of various courts but, as the Webbs (1908) made clear in their pioneering survey, no one model fitted all places.

At the time of the industrial revolution in Europe, therefore, the main urban settlements had separate administrative status – which brought with it certain rights, such as the election of members of Parliament in Britain. As urbanisation increased rapidly, so the system of urban government was modified to cater for the establishment of new industrial towns and the expansion of existing centres (Freeman, 1968). And at the same time, the incorporation of new lands outside Europe into the colonial systems of the industrialising countries included the introduction of similar separate forms of urban

128

government. Thus by the twentieth century, most of those parts of the earth's surface that were both permanently settled and within the orbit of the North Atlantic capitalist nations had a local government system which separated town from country.

Each country introduced its own peculiarities within this general pattern. Nevertheless, two main types can be identified. The first are in countries where central control of local government is relatively strong, and local administration is very much subservient to national ends. The other are in countries where local government enjoys a greater deal of autonomy. These two types are characterised by the United Kingdom and the United States of America respectively, and most of the discussion in this chapter will focus on those two countries not as stereotypes but as general examples of the two major forms.

Local Government Functions

The usual rationale for local government, serving an area which is but a small part of a sovereign state, relates to perceived efficiency and democracy. Regarding efficiency, it is argued that the local provision of services, controlled locally and accountable to the local electorate, ensures that services of the kind, quality and quantity required there are indeed provided; a central administration, perhaps with local representatives but accountable only to a central government, could not respond as readily to spatial variations in demand. Regarding democracy, it is argued that separate local governments provide a sensible counter to the potential of a centralised autocracy, and allow a sizeable proportion of the population to be involved in the government of the state. Local government thus advances efficiency, accountability, and democracy.

Despite these arguments, local governments have always been subsidiary to superior governments, for the simple reason that the latter are sovereign but the former are not. The existence of local governments is entirely dependent on central government support. Furthermore, the functions which local governments can and cannot undertake are usually specified — either in a constitution or in legislation passed by the sovereign parliament (or comparable institution); some autonomy may be allowed local governments in certain policy areas, but this is generally relatively slight (especially with regard to expenditure) and local autonomy is closely constrained. Thus although the initial demand for local government, especially urban local government, came from communities, the systems put into operation reflect the requirements of the central state — in varying degrees, as indicated in the UK/USA comparisons here.

Local governments were established to perform certain functions — the administration of justice, the relief of poverty etc. — required by the central state but which were best supervised by local residents (increasingly, elected local residents). For urban places there were, in addition, functions that were particular to such settlements, such as the administration of markets. From the outset, therefore, towns and cities had particular needs, which were reflected in the granting of separate status and powers. The nature of those needs changed during the period of rapid industrialisation and urbanisation that began in the nineteenth century, and led to even more particular forms of urban local government.

In general, larger settlements make greater demands on local government (or on government in general, which is provided locally) than do smaller places and rural areas. (There are exceptions to this; the provision of roads and other public utilities — water, electricity etc. — is usually most expensive, relative to demands, in thinly populated rural areas.) The reasons for this, according to Sharpe (1981) are:

(1) Urban populations are less self-sufficient. Rural residents have much less need for public recreational facilities than do their urban counterparts; they can burn their own rubbish, too, which innercity residents cannot. Thus in urban areas the state must provide — or at least oversee the provision of — facilities that elsewhere are provided by individual residents.

(2) Urban population densities are much greater than those of small settlements and rural areas, which means that there are many more negative externalities created in towns and cities. Negative externalities are costs imposed on people by others, over which the former have no control. Some of these costs may be in the form of potential hazards to human health and life, as in the likely effects of the discharge of untreated sewage into water supplies. To control such externalities local governments in urban areas take on a range of regulatory functions. Furthermore, the density of land use is such, and has increased so, that management of interaction within the city, through traffic planning, is a necessity for urban governments.

(3) Urban settlements, especially large and dense ones, generate needs that are generally absent elsewhere. Paved footpaths, street lights and extensive policing are examples of the facilities and services

generally considered necessary in urban areas. It is not necessary that they be provided by the state, and in some countries — and even more in parts of some urban areas of many countries — they are either provided collectively by the local residents independent of the state or provided by private sector firms. But in most cases, local government provisions of such services and facilities is considered the only feasible method.

(4) Urban areas, especially in their inner parts, contain considerable obsolescent physical plant, the renewal of which is uneconomic to the private sector. Thus local government is required to participate in the maintenance and renewal of the urban fabric, in ways that are not relevant outside the large urban areas.

According to this argument, urban local governments are involved in two broad types of function. The first comprises those functions that are provided by all local governments, urban or not, such as education and social services; it may be, however, that the demand is greater per capita in urban than in non-urban areas, and perhaps the costs of provision are greater (wages tend to be higher in large cities, for example, and land dearer), so that the relationship between size and local government activity is non-linear; the costs of running urban government per capita increase as urban size increases. Secondly, there are those functions that are peculiar to urban places, and reflect their size and density. To these must be added those that reflect the position of cities, especially large ones, in the society of which they are part; the provision of museums and art galleries, concert halls and sports centres, and international airports is a local function in the large cities of some countries, as is the servicing of large shopping centres, and such provision almost invariably is subsidised from local revenues.

The above discussion sees the state — which creates and directs local governments — as an institution which responds to local demands. While this is undoubtedly so to a certain extent, other analyses suggest that the state, including its local government, has a much more central role in the economy of capitalist societies, especially late (or advanced) capitalist societies (see, for example, Clark and Dear, 1981, and Johnston, 1982). In this alternative conception, the role of the state in capitalist society is presented as maintaining the structure and relationships of the local social formation; it must provide the environment (physical and social) for economic success, and it must ensure consensus support for the economic system. Its operations and expenditures are directed towards these two goals of accumulation and legitimation.

With regard to the former, it is involved in expenditure on social capital, which can be divided into: (1) *social investment* aimed at increasing the productivity of labour and the rate of profit; and (2) *social consumption*, which involves subsidising the costs of labour to the private sector. Regarding legitimation, it is involved in, (3) *social expenses* on maintaining peace and harmony within society (through welfare payments, for example, and in the provision of police forces). The functions of the state, according to this view, are to facilitate and subsidise capitalist accumulation and to ensure the population accept this method of organising society. Local governments are part of this operation, and their actions as described above can readily be fitted into the framework provided here (based on the work of O'Connor, 1973).

The Pattern of Urban Local Government

Given the above description of local government functions, plus an ability to identify criteria by which such concepts as efficiency can be operationalised, it should be possible to identify the optimum pattern of local government (in much the same way as central place theory suggests an optimum settlement pattern). But the system of local government has not been created and re-created in the search for such optima; instead it has evolved in what is very much an *ad hoc* manner, overseen by the central state.

Urban Local Government in Great Britain

In the early nineteenth century, at the beginning of the industrial revolution, British local government comprised the boroughs and the shires, with the latter subdivided into hundreds or similar administrative units. As industrialisation and urbanisation proceeded so did the demands for more local government, especially in urban areas — including those lacking borough status because of their recent growth. The problems of public health were perhaps the most crucial in these large, dense concentrations of relatively poor individuals, and administrative units — urban sanitary districts — were created to minimise the negative externalities that were a concomitant to urban growth. (And were not new to it in the nineteenth century, as illustrated by the desire of Elizabeth I to limit the growth of London in the sixteenth century.)

The creation of an administrative system to supervise public health in urban areas was undertaken by the central government, and the local governments were subject to the relevant Acts of Parliament.

Other functions were allocated to these units (by the Public Health Act of 1875, for example), and at the end of the century a new system was created. Under the 1888 Local Government Act, *county borough* status was granted to the largest urban places – all those with populations exceeding 50,000 plus thirteen other smaller places (such as Canterbury) which were thought to justify that status. These county boroughs were independent of and equivalent to the counties which surrounded them, having all of the powers also allocated to the latter. Thus big city government was separated from that of the surrounding areas – creating 'urban islands' – in recognition of the particular problems of such agglomerations.

Within the counties, two types of separate but not independent authorities were created to govern the smaller urban places. *Urban districts* and *municipal boroughs* differed in their size (the latter were the larger, with a median population in 1927 of just over 10,000 compared to just over 5,000 for the former) and in the number of functions allocated to them. Again, this reflected an appreciation of the separate problems of urban places, especially the larger ones; certain functions were retained by the encompassing counties, notably education, because the smaller places could not provide an effective separate service. Finally the County of London, comprising an entirely built-over area by 1888, was subdivided into *metropolitan boroughs*, subsidiary units which evolved from the ecclesiastical parishes, poor law commissions, turnpike boards and other *ad hoc* bodies, most of them established during the nineteenth century to solve increasingly pressing problems stimulated by rapid urban expansion.

The rapidity of urbanisation and urban expansion in the twentieth century meant that the administrative structure established for urban Britain in 1888 was soon obsolescent. This was particularly the case in the industrial heartlands and around the national capital. In the former, where the creation of conurbations as cities grew and met and the expanding built-up area reached the boundaries of formerly separate small market towns, there was little correlation between the urbanised area and the provision of urban government. Each of the major conurbations in 1951 contained a large number of separate local government units (Table 5.1). As the built-up area expanded, so some of these units, especially the larger ones, were able to extend their territories, annexing land formerly under county jurisdiction in rural districts. And in some cases, formerly separate units were merged into their neighbours.

The complexity of urban government in Britain was a cause of much

Table 5.1: The Local Governments of British Conurbations, 1951

	Greater London	Southeast Lancashire	West Midlands	West Yorkshire	Merseyside	Tyneside	Central Clydeside
Population	8,348,023	2,422,650	2,237,095	1,692,687	1,382,443	835,533	1,758,454
Number of:							
Metropolitan Boroughs	29	—	—	—	—	—	—
County Boroughs	3	7	6	6	4	4	1
Municipal Boroughs	36	15	8	6	2	2	—
Urban Districts	26	29	10	21	6	7	12
Rural Districts	1	2	0	0	0	0	7
Average Population		45,710	93,212	51,293	115,203	64,272	
Average Population excluding CBs		20,957	34,461	20,510	34,367	28,401	

concern to administrators and others, so much so that certain functions were taken away from local government and allocated to *ad hoc* authorities covering larger areas and which were accountable only to the central government. (Some had local nominees but not local representatives on their governing bodies.) Attempts were made to rationalise the system, with Local Government Commissions appointed in 1927 and 1958, but although some changes were made there was no imposition of a structure which equated each built-up area with a single urban government (as Table 5.1 shows). And yet to many observers the need for such a structure was increasing, in particular because of the problems of planning for growth — including that of traffic — and for renewal in the obsolescent parts of the urban areas.

The most recent attempts to tackle this problem began in the late 1950s with the establishment of a Royal Commission on the Government of Greater London. According to the 1951 Census, the metropolitan boroughs of the London County Council area had a population of 3.3 million whereas the entire Greater London conurbation (the Metropolitan Police District) contained 8,348,023; outside the LCC there were three county boroughs, 36 municipal boroughs, 26 urban districts and one rural district. To provide a more rational system of local government, a new Greater London County was created in 1964; it was divided into 32 metropolitan boroughs, which were independent units for many functions, including education. Even this was considered insufficient by some, for the new GLC did not embrace the entire 1951 built-up area, with consequent problems for the strategic planning responsibilities that it was allocated.

For the remainder of the country separate Royal Commissions were established in 1966 for England and Wales and for Scotland, again with the brief of suggesting a more rational system of local government. The arguments favoured by the Commissions were those of the economies of scale — bigger is more efficient — and of the problems of town-hinterland administrative separation. Thus they proposed a series of 58 single-tier counties for England and Wales, plus 3 two-tier metropolitan counties: town and country were to be integrated in large, multi-purpose authorities. (Ten regions were suggested for Scotland, with a two-tier structure.) But this solution was unacceptable politically to the Conservative government elected in 1970, and a new system was introduced for England and Wales (in the 1973 Local Government Act) which retained a two-tier structure throughout the country. In the six metropolitan counties, most of the power was allocated to the lower tier so that, for example, Greater Manchester comprises eight quasi-independent metropolitan districts. Elsewhere, much of the power was allocated to the upper-tier authorities (the counties); only the largest towns had separate administrative units (county districts), with control over few important functions.

Most of the functions of British urban local governments relate to social consumption and social expenses. The largest item of expenditure (about half of the total), for example, is education, which is provided as part of the production of a labour force with the requisite skills; expenditure on public housing is also in that category, whereas that on police forces and fire brigades falls under social expenses. To perform these functions, the local government structure evolved in an *ad hoc* manner, with few urban jurisdictions – especially the larger ones – corresponding to the functional urban areas, let alone those areas plus their hinterlands. Several attempts to reform this structure have been made by the central government. The guiding principle of the reforms has been 'bigger is better'. Implementation has rarely followed this through, however, because of the conflict between the efficiency criterion and that of local democracy and accountability, which latter calls for small areas, and also because of partisan desires among reformers to protect their electoral strongholds (see, for example, Johnston, 1979, and Taylor, 1983). The present system is thus a compromise between various interests – and one which seems to please none of them.

Urban Local Government in the United States

Several aspects of the American local government system reflect their origins in British practice; the main difference is the absence of any strong central control and implementation of reforms. As a consequence, in most urban areas local government is much more complex and fragmented than in Britain. (It should be noted here that in the American context central government relates to the State governments. Local governments are not mentioned in the US Constitution and so the Federal government has no control over them. They exist as creations of State legislatures, and are protected – in many cases – by being mentioned in State Constitutions.)

As in Britain, the basic element of local government in nearly every State is the *county*. Within this, *municipalities* (the name varies between the States) can be incorporated to govern defined areas, if they meet threshold requirements of population size and density. Incorporation generally follows a petition to the State legislature from the residents of particular areas and a referendum of all residents. Once incorporated, a municipality has certain statutory powers as defined by either the Constitution or, more often, State legislation; in most States there are also constraints on, for example, how income can be raised, and in what quantities.

When most of the country was settled the urban areas were small and municipalities were incorporated to cover the entire built-up area. As the latter expanded — especially with the suburban boom from the late nineteenth century on — it spilt over the municipal boundaries into the surrounding counties (what is generally known as unincorporated territory); in some cases formerly separate municipalities were engulfed by the suburban sprawl of a larger neighbour. Some new municipalities were incorporated in these suburban extensions, but it was general in the nineteenth and early twentieth centuries for the newly-settled suburbs to be annexed to the existing municipality covering the older parts of the built-up area. This was of value to the government of that municipality, since it extended its tax base (real property values — see below); it also benefited the residents of the new suburbs, because the existing municipality was better able to provide basic utilities and other facilities than either a county government or a newly-incorporated municipality (Teaford, 1979).

Up until about the end of the nineteenth century, therefore, American municipalities expanded in area to encapsulate the newly-developing suburbs (in several cases, large tracts of undeveloped land were annexed in anticipation of suburbanisation). But since then, the pace of annexation has slackened substantially, especially in the longer-settled parts of the country. Instead, residents of new residential developments — and also the developers of industrial and commercial complexes in the suburbs (Johnston, 1981a) — have preferred to incorporate separate suburban municipalities, most of them small. Such administrative separation brings several advantages to suburbanities.

(1) Because local governments raise their taxes on local property values, the residents of a suburban municipality must pay taxes only for the operation of their own district and need not contribute to the costs of administering, and providing services for, other parts of the urban area. Since most suburban residents are relatively affluent, this means that the rich can avoid subsidising social consumption for the poor.

(2) Because local governments are independent, residents of small suburban muncipalities can avoid the political activities of larger places — in which they might constitute a small minority

unable to obtain desired benefits for their home districts.

(3) Because municipalities are independent land use planning authorities in most States, residents of small suburban municipalities can manipulate the zoning scheme to exclude undesirable land uses and users and so control the social and physical fabric much more so than is the case in larger places (with consequences for control of the educational environment — see below).

In effect, therefore, the relative autonomy of municipal governments, and the ease of incorporation in most States, encourages the residents of unincorporated county areas to create separate suburban muncipalities to protect their own interests rather than to join the muncipality representing the core of the urban area, in which local interests would be submerged in wider political conflicts. (See Johnston, 1981b.)

The growth of the incorporation movement means that, as a broad generalisation, American metropolitan areas can be divided into two zones —an inner one comprising the core of the built-up area and governed by a single municipality (known as the central city); and an outer mosaic of separate municipalities (the suburbs). According to the 1977 Census of Governments, there were 18,862 municipalities in the country, 6,444 of which were in the 272 Standard Metropolitan Statistical Areas (SMSAs). (The number of municipalities within SMSAs increased by over 800 in the five-year period between the 1972 and 1977 censuses.) Most of these were small, with 3,241 having populations of less than 2,500 and only 391 (including the central city of most of the SMSAs) having more than 50,000. The Chicago SMSA alone had 261 municipalities (Table 5.2) and San Francisco-Oakland had 59, on average much larger than those in Chicago.

This administrative separation of the various parts of suburbia, both from the central city and from each other, is used in particular by the relatively affluent to accentuate the differences between rich and poor and to keep the two (and thus, implicitly, the rich whites and the poor blacks) apart. The fragmentation is magnified in most States by the existence of other local government institutions, notably the *school districts*. When public education systems were introduced, responsibility for its provision was handed by the States to the county governments and to the large municipalities. In most States (the main exceptions were in the South where county governments retained control) this responsibility was devolved from the county level to individual school districts, which enjoyed a large degree of autonomy and were financially independent, relying for revenue on taxes on local

Table 5.2: The Local Governments of the Chicago and San Francisco-Oakland SMSAs, 1977

	Chicago	San Francisco-Oakland
Municipalities (by population)		
50,000+	12	12
25,000—49,999	32	15
10,000—24,999	67	15
5,000—9,999	42	9
2,500—4,999	41	5
1,000—2,499	33	1
1—999	34	2
School Districts (by pupils)		
100,000+	1	0
50,000—99,999	1	2
25,000—49,999	1	6
12,000—24,999	7	6
6,000—11,999	33	20
3,000—5,999	50	13
1,200—2,999	110	17
300—1,199	96	15
1—299	16	9
Special Districts	519	147

property. Thus prior to suburbanisation, the counties surrounding most cities had a number of school districts serving the rural population. Most were small and operated only a few schools; in some cases elementary and high school education were provided in separate, not necessarily overlapping districts, whereas in others there were single 'unified' districts.

When the territories of these rural school districts were invaded by expanding urban areas, the fate of their separate status depended on the decision regarding annexation. If the areas were annexed into cities which provided an educational service, then the school districts were dissolved. But if the areas either remained as unincorporated county land or were incorporated into separate municipalities, the school districts were retained. Their separate existence was an extra attraction to affluent suburbanites, who paid through their property taxes to support local schools only and who could, through control of the zoning power in the municipalities, ensure that these local schools did not contain many poor and/or black students, who are widely perceived as social negative externalities (Johnston, 1981a).

In all, there were 15,174 school districts in the country in 1977,

5,520 of them within the SMSAs; as Table 5.2 shows, Chicago and San Francisco-Oakland both had substantial numbers of such districts, most of them small. In Chicago, 75 of the districts operated one school only, and only 8 contained more than 20; the comparable figures for San Francisco-Oakland were 12 and 11.

The mosaic of local government units in American metropolitan areas is completed by the *special districts*. These are *ad hoc*, single purpose authorities, established to perform a particular function for a territory that may coincide with those of no other units; some special districts cover entire metropolitan areas. Some of these are remnants of a pre-metropolitan situation, such as the volunteer fire districts which service some suburban rings (the Chicago SMSA had 146 of these in 1977). Others have been established for a variety of reasons, including (Stetzer, 1975): constitutional limitations on municipal revenue-raising, which make separate districts the only way of financing certain desired services; constitutional prohibition of differential taxing within a district, so that payment for a service matches its use; a desire to finance services through particular user charges rather than general taxation; and a requirement for certain Federal grants that particular functions be undertaken (such as regional planning). The special district has become increasingly popular as a means of avoiding the existing system, and their number increased from 8,299 in 1942 to 25,962 in 1977; 9,580 of the latter were in SMSAs, including 519 in Chicago and 147 in San Francisco-Oakland.

The multiplicity of local government units in American urban areas, particularly the suburbs, is a key element in not only the political but also the economic and social geography (Johnston, 1981a). The availability of these areas has been employed by residents to manipulate the geography of suburbia to their social and fiscal advantage – they avoid contributing through local taxes to the expenditure on social consumption and social expenses for other groups in the metropolitan area (generally the relatively poor, who live in the central cities) and can ensure that only what they want is provided – as far as possible, for themselves alone. This administrative fragmentation is much greater in most SMSAs than it has ever been in comparable British urban areas. It has survived because State governments have been unprepared to use their powers to impose basic changes on the suburban way of life.

Elected Governments and Public Officials

As in almost all democratic countries, local governments in Britain and the United States are run by a combination of elected politicians and appointed public servants. The relative power of these two groups is a subject of considerable academic debate at the present time (see Saunders, 1979, and Dunleavy, 1980). Many argue that most of the real power lies with the appointed officials who, through their professional norms and expertise and their ability to manipulate agendas, are better able to control the activity of a local government than the, usually part-time, elected representatives.

In Britain, despite the managerial power of the appointed officials (see Johnston, 1982), it is clear that the elected representatives have some considerable influence on the general format of local policy. In recent decades, the national political parties have dominated elections to most district and county councils, and only in a few rural areas are large numbers of independents still elected. (Most of these are politically inclined towards the policies of the Conservative party.) Of the two main parties, where Labour is in control there is generally above-average expenditure on social consumption (particularly education) and certain social expenses (notably social welfare), *ceteris paribus*, whereas in Conservative-controlled authorities there is greater than average expenditure on those social expenses (such as police) concerned with the protection of private property. In this way, political parties use their control of local governments to benefit the interest groups within society from which they obtain the majority of their support.

In the United States, this manipulation of local government to serve vested interest groups takes a number of forms, many of which have no relationship to the country's two main political parties. During the late nineteenth and early twentieth centuries, the politics of many of the large cities were controlled by a combination of business and immigrant group interests; graft and corruption were common. This was not approved by others, especially the middle classes, who believed that local government should be apolitical; municipalities should be run like businesses. To further this view, a reform movement was launched, and the National Municipal League published a model municipal charter in 1899 (Judd, 1979). To overcome the 'machine' politics of the big cities, this proposed at-large rather than ward elections, to prevent minority representation and porkbarrel politics, and the abolition of party tickets on ballot papers. Further, it suggested weakening the power of the elected Council and its replacement by either a strong mayor (who could veto the council and who appointed the major public servants) or a small elected commission who supervised the running of the muncipality by a city manager and his appointed officers. Although the reform platform has been adopted in some of the largest cities, many of its successes have been in the suburbs, where apolitical, business-like government favours the affluent middle class who wish to keep taxes down and avoid subsidising the poor.

A major feature of local government for much of the time is voter apathy; the electoral turnout tends to be very low. This can reflect a variety of circumstances; it could, for example, indicate either satisfaction or a belief that change is impossible. In Britain, local elections tend to mirror national political trends, with the unpopular party in the national parliament (almost invariably the government in the recent economic crises) losing support — usually through abstentions rather than defections. Occasionally, local issues attract considerable attention, especially in those American States where issues can be put to the vote at a referendum if sufficient electors sign a petition; the tax-cutting Proposition 13 approved by Californian voters in 1978 is an excellent example of this. (Introduction of the referendum, and the popular initiative leading to it, was part of the reform programme: Butler and Ranney, 1978.) Such events are relatively infrequent, however, and in general local government attracts little interest — although its bills often create much voter discontent.

Financing Local Government

As indicated in several places in this chapter, a major source of revenue for local governments is a tax on local property values (known as the property tax in the USA, where it is based on assessed market values, and as rates in Britain, where the base is an imputed rental value). Because the territories of various local government areas differ in the value of their land — both within and between urban areas — the basic fiscal resource available to local governments varies substantially, especially if there is no other major source of revenue constitutionally available (local governments in Britain cannot raise income taxes, for example). The result is a complex geography of public finance (Bennett, 1980); some local governments have substantial resources relative to both their needs and their inclinations to spend, but others do not (Johnston, 1979).

In recent years, the demands on local government have increased substantially. In part, it is argued (Newton, 1980), this is because of the nature of the services it provides. There is, for example, the need to increase the quantity of local services in the face of certain changes; each generation seems to demand more education, for more of its children and over a longer period of time, than did the last; more cars on the roads means more traffic control schemes; and so on. And, of course, there is a general desire for a better quality of service — not just a swimming pool but one of Olympic standard. New services are introduced — environmental protection, for example, and provisions for the handicapped. Obsolete physical capital has to be replaced. And, especially in times of inflation, the costs of such services increase very rapidly, because they are labour-intensive.

At the same time as these demands for more and better services have been received, so local governments in many parts of the world find it difficult to finance them. Newton (1980) suggests three reasons for this. First, their tax base is not very suitable. With income taxes, for example, as incomes rise so does the tax take. But with property taxes, the income does not increase unless properties are regularly revalued. Thus property taxes and rates have to be increased regularly, which creates a further problem because of their visibility; whereas most income taxes are paid through a scheme (PAYE) which means that the taxpayer never receives the money, property taxes are separately demanded, in many cases as a fixed sum. Finally, because the payment of property taxes is the main contact between most electors and local governments, such taxes are politically visible and can generate electoral protest.

Local governments are experiencing a fiscal squeeze, therefore, especially those forced to rely heavily on property taxes (Newton, 1980; Sharpe, 1981). Furthermore, their ability to raise revenue sufficient to meet their expenditure requirements can vary substantially. Property values are generally highest in areas with buoyant economies but the demands for certain types of social expenses are frequently greatest where the economy is in relative, if not absolute, decline. Thus the government of an urban area in a prosperous region is probably better able to raise property taxes and finance expenditure than is one in a declining area. As a result, differences in the quality of life between such places tend to be exaggerated.

To counter such differences, the central government in the UK has for several decades provided a substantial amount of financial aid to local governments. Known now as the Rate Support Grant (RSG) this was structured until 1981 so that there was a negative relationship between a place's fiscal resources (rateable values per capita) and grant volume and a positive relationship between its needs (the demand for services) and its grant (see Bennett, 1980, 1982a, 1982b). In the late 1970s, the RSG contributed some 61 per cent of the total income of local governments. Nevertheless, rate levels in most local government areas were considered high and, for reasons outlined above, were increasing rapidly. To reduce local government spending in total, the government altered the RSG in 1981 as part of its general fiscal policy. A target expenditure for each area is determined, relative to its centrally perceived needs, and a grant towards that expenditure is provided. If the local government seeks to spend more than the target, it is penalised — by withdrawal of all or part of RSG. In this way, local autonomy is being considerably reduced and the geography of local spending reflects the desires of central government.

In the USA, the inter-urban differences in needs and resources similarly create substantial variations in the ability of local governments to meet demands, and Federal and State programmes have been devised to redistribute public money towards areas of greatest need (Bennett, 1980). The variations have been accentuated in many places by the city-suburban fragmentation, which means that the central cities are the homes of the most needy but, increasingly as industry and commerce flee to the suburbs, they have the poorest per capita fiscal resources. The mismatch of needs and resources has led (in conjunction with other factors: David and Kantor, 1979) to financial crises (in a few places near-bankruptcy) in some central cities, in comparison to the relative prosperity of many suburban governments.

Increasingly in recent years observers have talked about an urban fiscal crisis, as part of a general fiscal crisis of the state. Some argue that this is overly dramatic, that local governments are suffering a resource squeeze, but are still surviving. (See Sharpe, 1981, and Newton, 1980, both of whom argue that the 'fiscal crisis' is most apparent in countries where local government relies heavily on property taxes.) Certainly, bankruptcies and an inability to meet bills are not (yet) common. But many argue that increasingly local governments cannot meet the demands made on them and that as their budgets become more constrained by central government policy this is likely to continue.

Conclusions

In Western Europe at least, local government preceded the present pattern of central governments, in that the current major local government units — such as the shire counties in Great Britain — were at one stage quasi-independent states which were later incorporated into larger units and made subsidiary to them. Urban local government has almost invariably always been subsidiary to some larger unit, however, and its nature and operations have been largely shaped from above. Most urban local governments date back no further than the late nineteenth century in the advanced industrial countries, by which time strong central states had been established and local urban autonomy was relatively weak. Nevertheless, as demonstrated in the comparison here, the nature and degree of that autonomy — and the degree to which it was challenged from the centre, varied substantially from country to country.

Urban local government, therefore, must be studied as part of the state apparatus which exists to advance the twin goals of a capitalist society — accumulation and legitimation. Of the three main expenditure functions of the state (see above, p. 132), local governments have been allocated important roles in social consumption and social expenses. Most of their activities and spending focus on the reproduction of the labour force and the maintenance of social harmony (Dear, 1981), and although some local activities — aspects of land use and transport planning, for example — can be categorised as social capital expenditure, this remains largely the preserve of the central government.

Although, as illustrated here, the nature of the urban local government system varies substantially between countries, there are many shared elements; the so-called present 'urban fiscal crisis' is one of them. This is because of the role of the state in modern society. To protect firms from the problems of economic crisis, the state has taken an increasing role in subsidising the reproduction of the labour force, hence an increase in the demands for social consumption expenditure (including what is known as collective consumption expenditure on public goods). At the same time, to maintain social harmony in crises there has been an increased demand for social expenses. The result is increased state spending, which means increased borrowing and higher taxes, both of which are considered to inhibit economic growth. To achieve the latter, the state needs to cut its social consumption and expenses outlays, it is argued, while increasing its social capital expenditure to encourage business succcess. The state is in a fiscal squeeze; local government is strongly affected by this.

Many consequences follow from the increasing fiscal squeeze (which may be a permanent feature of late capitalism). One is the removal of functions from local governments, reducing local autonomy and variability in the interests of central economic guidance. Another is the restructuring of local government, to make it more efficient (in the same way that industry is restructured). As the state pursues its twin goals, therefore, they come into conflict. Encouraging accumulation means promoting efficiency, which in local government means reducing democracy and (potentially) harming legitimation. Because they have been allocated a central role in the maintenance of social harmony, urban local governments may (if they are allowed to remain) be the focus of the efficiency and democracy battle of the future.

References

Bennett, R.J. (1980) *The Geography of Public Finance*, Methuen, London

Bennett, R.J. (1982a) *Central Grants to Local Governments*, Methuen, London

Bennett, R.J. (1982b) 'The Rate Support Grant in England and Wales' in D.T. Herbert and R.J. Johnston (eds.) *Geography and the Urban Environment*, vol. 4, John Wiley, Chichester

Butler, D.E. and Ranney, A. (1978) (eds.) *Referendums*, American Enterprise Institute, Washington

Clark, G.L. and Dean, M. (1981) 'The State in Capitalism and the Capitalist State' in M. Dear and A.J. Scott (eds.) *Urbanization and Urban Planning in Capitalist Society*, Methuen, London, 45–62

David, S.M. and Kantor, P. (1979) 'Political Theory and Transformations in Urban Budgetary Arenas' in D.R. Marshall (ed.) *Urban Policy Making*, Sage Publications, Beverly Hills, 183–222

Dear, M. (1981) 'A Theory of the Local State' in A.D. Burnett and P.J. Taylor (eds.) *Political Studies from Spatial Perspectives*, John Wiley, Chichester, 183–200

Dunleavy, P. (1980) *Urban Political Analysis*, Macmillan, London

Freeman, T.W. (1966) *The Conurbations of Great Britain*, University of Manchester Press, Manchester

Freeman, T.W. (1968) *Geography and Regional Administration*, Hutchinson, London

Hechter, M. and Brustein, W. (1980) 'Regional Modes of Production and Patterns of State Formation in Western Europe', *American Journal of Sociology*, 85, 1061–94

Johnston, R.J. (1979) *Political, Electoral and Spatial Systems*, Oxford University Press, Oxford

Johnston, R.J. (1981a) 'The Political Element in Suburbia', *Geography*, 66, 286–96

Johnston, R.J. (1981b) 'Local Government Suburban Segregation and Litigation in US Metropolitan Areas', *Journal of American Studies*, 15, 211–30

Johnston, R.J. (1982) *Geography and the State*, Macmillan, London

Judd, D.R. (1979) *The Politics of American Cities*, Little Brown, Boston

Newton, K. (1980) *Balancing the Books*, Sage Publications, London

O'Connor, J. (1973) *The Fiscal Crisis of the State*, St Martin's Press, New York

Saunders, P. (1979) *Urban Politics: A Sociological Interpretation*, Penguin, London

Sharpe, L.J. (1981) (ed.) *The Local Fiscal Crisis in Western Europe*, Sage Publications, London

Stetzer, D.F. (1975) *Special Districts in Cook County*, University of Chicago, Department of Geography Research Paper 169, Chicago

Taylor, P.J. (1983) 'The Changing Political Map' in R.J. Johnston and J.C. Doornkamp (eds.) *The Changing Geography of the United Kingdom*, Methuen, London

Teaford, J.C. (1979) *City and Suburb*, John Hopkins University, Baltimore

Webb, S. and Webb, B. (1908) *English Local Government vol. III*, Longman, London

6 RETAILING

R.L. Davies

Introduction

This essay takes both a forward and backward look at retailing in the urban context. It seeks to combine a review of certain sections of the literature that have dealt with recent changes in retailing with a series of questions about what will or should be the dominant concerns of geographical enquiry in the future. The one does not necessarily follow from the other, however. Rather the essay takes as its terms of reference the fact that the retail industry stands at the present time at a crossroads. The economic recession has coincided with what appear to be some fundamental shifts in the climate of retailing, in its background market conditions, its potential for innovation and development and in the planning context in which retailing operates.

These changed circumstances are not in themselves unique to retailing. They will have considerable implications for the course of events and nature of research in the manufacturing, office, transport and other fields. They are perhaps more prominent in the retailing case, however, by virtue of the extreme sensitivity of this industry to what happens within the general economy and in the attitudes and aspirations of society. At the present time we can look back to an era of retailing that was largely exemplified by growth, that between about 1960 and the late 1970s was characterised by the introduction of new, large-scale activities and which was accompanied by a radical transformation in shopping behaviour, most notably the switch from daily shopping trips to once-weekly expeditions for grocery goods and comparison shopping for durables on a more infrequent basis. In the 1980s, we can expect some continuation or consolidation of these trends; but there have clearly emerged some limits to further growth, a new set of considerations regarding the forms of future development, and a great deal of uncertainty about how consumers will react to different kinds of opportunities from those presented hitherto.

We will examine first what the new dimensions to retailing seem to be. Secondly, the themes of recent past research will be identified, particularly from the standpoint of the traditional forms of geographical enquiry into the spatial structure of the physical environment. Finally, an attempt will be made to link up the course of this research to those questions that will need to be addressed given the future state of the retail system.

The New Dimensions to Retailing

The Changing Market

Prior to the current recession, retailing in Britain had already been forced to anticipate profound changes in the demand for goods and services from those demographic and economic conditions that emerged in the second half of the 1970s. On the one hand, the population of the country appeared to stop growing due to the dramatic decline in the birth rate; on the other hand, real disposable income also became depressed due to a combination of hyper-inflation, growing unemployment and the effects of the oil crises on petrol costs. Per capital spending on convenience, mainly food goods, dropped markedly and expenditure on comparison, mainly durable goods, became static (Unit for Retail Planning Information, 1980a). These portends concealed a number of cyclical processes at work, however, the effects of which are likely to be more beneficial in the short term rather than long term.

On the demographic side, we know that there has recently been an increase in the birth rate which, coupled with the continuation of high life expectancy rates, can be expected to lead to a temporary expansion of the population in the next few years. Thereafter, nevertheless, we can assume that the birth rate will again decline and, in the absence of any major alteration to levels of mortality, the total population will once more contract or become static. More significant than these fluctuations in population numbers during the 1980s, however, will be the changes in the composition of the population in terms of age groups and household size categories. We can be certain that the proportion of the elderly will increase and that, with the trend to later marriages and higher divorce rates, there will be a lowering of the mean household size.

The economic outlook is less clear but we might again expect some improvement sooner rather than later. In the short term, we can assume that there will be some partial recovery from the world-wide recession, a continued reduction in inflation, some easing of the present high levels of unemployment and a small reduction in taxation (Henley Centre for Forecasting, 1982). All this will lead to a better standard of living for the majority of the population, generating greater amounts of spending and giving a boost to retail turnover. The improvement should not be exaggerated, however, for much of the inherent growth will constitute a catching-up on past shortfalls and there may well be some unforeseen events which have a substantial negative effect. In the longer term, we can be fairly sure of there being further international crises affecting in particular both energy reserves and monetary flows. These, together with the cumulative impact of new technologies on work practices, stronger competition from newly developed countries in the production of consumer goods and inevitable fluctuations in interest and inflation rates are likely to lead to a retraction of the earlier gains.

These rather dismal prospects for the nation as a whole need to be looked at alongside the shifts likely to take place in more localised urban markets. The results of the 1981 Census of Population have already confirmed that the trend to increased suburbanisation and inner city decline that was so prominent in the 1960s was even more pronounced during the 1970s. This has had two major repercussions. First, there has been an increase in the size and status of smaller settlements in the urban hierarchy at the expense of much larger places (Randolph and Robert, 1981). Secondly, the concentration of the older, poorer and less mobile sections of the population in inner city areas has become intensified. Despite the promise of gentrification, the formation of smaller households and an increase in the diseconomies of commuting, it is unlikely that we will see a significant abatement of these processes during the next decade. The picture for the end of the 1980s, therefore, is one of a diminution of the effective purchasing power of the population as a whole but with a further dispersal of the main reservoirs of consumption to the peripheral parts of the urban area.

Innovations and Development

The major expansionary phase of retailing between 1960 and 1975 was of benefit primarily to the large corporate chains. During this period, there was a dramatic decline in the total number of shops but a vast increase in the absolute and proportional amounts of trade commanded by the multiple groups. There were two main agencies that fostered this growth: the large supermarket and subsequently its metamorphosis into superstores and hypermarkets; and the purpose-built, managed shopping centre which provided additional floorspace in the comparison or durable-goods trades.

There are surprisingly few statistics regarding the explosive increase in large supermarkets throughout the country in the 1960s, but it was these rather than the later superstores or hypermarkets that initiated the major changes in food shopping behaviour and which in turn were primarily responsible for the decline of traditional grocery shops. Superstores and hypermarkets were strongly resisted by local authorities and central government until the middle part of the 1970s. Only 10 of these developements were opened during the 1960s, a further 52 in the first half of the 1970s; but 106 followed between 1976 and 1979 (URPI, 1980b). Clearly, the new market conditions that appeared in the second half of the 1970s have had no dampening effect on their growth and indeed most of the multiple firms engaged in their development have ambitious expansion programmes for the next few years.

The types of consumers to which superstores and hypermarkets are oriented, however, are precisely those identified at the end of the last section as becoming more entrenched within outlying areas. Not unexpectedly, therefore, the outer suburbs are also the most preferred location for their development although several have been opened in town centres and even within the inner city (Pacione, 1979). The questions for the future are how much further growth of these stores can be supported in suburban localities and at what stage will a relative saturation set in?

The growth in the number of purpose-built shopping centres has followed a somewhat different course to that of the large grocery outlets. This is most clearly seen in the recent history of the town centre schemes, which have provided the greatest amount of new floorspace, but their fortunes have generally been paralleled by the smaller centres established in residential areas. Almost 100 town centre shopping schemes had been constructed in Britain by the end of the 1960s and a further 145 were opened in the first half of the 1970s (Davies and Bennison, 1979). The years 1974 to 1976 represented a peak in this development, however, since when there has been a relative decline. The earlier trend to increasing sizes of centres has also become reversed so that there are now proportionately more smaller sizes of schemes being built in smaller sizes of places (Hillier Parker, 1979 and 1981).

The future prospects for shopping centre development, at least in the larger central areas, would therefore seem to be poorer than those for superstores and hypermarkets. There are three factors that explain this. First, the new schemes were a clear response to the increased consumer spending of the 1960s and early 1970s and the subsequent market conditions that have emerged have deterred much further investment (Lee, forthcoming). Secondly, the relative erosion of the trading strength of the larger town and city centres, albeit masked by the introduction of new schemes, does not bode well for the creation of additional floorspace and would likely in any case be resisted by local planning departments. Thirdly, development costs within the larger central areas have escalated in recent years and have made new schemes less economically viable propositions.

There are differential growth prospects for a variety of other retail innovations and developments that emerged in the 1960s and 1970s, however. On the brighter side, we can expect to see an increase in such activities as discount retail warehouses, fast food outlets, DIY and garden centres and mini-markets or so-called modern convenience stores. These all tend to be oriented to car-borne shoppers again, nevertheless, and prefer locations that are either suburban based or aligned alongside major routeways. More negatively, the trend to greater diversification in the merchandising policies of the more traditional but larger high street stores (or what has been referred to as scrambled merchandising) suggests a growth in competition that may not be sustainable throughout the 1980s. Several observers in fact have predicted the collapse of certain of the major variety store groups towards the end of the decade which would exacerbate the decline of the larger town and city centres.

The Planning Context

The present time is particularly significant from the planning point of view for it marks a shift in attention away from the preparation of structure plans to the introduction of a new wave of district or local plans. In the first two years of the conservative government, we have also seen some relaxation of development controls which, whilst its significance for retailing has so far been minimal, suggests that there will be a more flexible approach to retail planning in the future. At the same time, there have already been clear signs that planning has become much more cautious, with the abandonment of long range forecasting and an emphasis instead on simply monitoring changes as they occur. This reflects again on an adaptation to the new market conditions to be found but also on what the likely retail response will be.

The structure plans of the 1970s look in hindsight to have been disappointing exercises in so far as their retail planning provisions are concerned (Guy, 1980). They were of course the product of a planning philosophy tuned up to growth. The strategic matters they mostly dealt with were therefore how much further retail floorspace could and should be accommodated within the county and where might any additional resources be best allocated in space. In many cases (especially the earlier plans), retail potential models were used to arrive at these estimates and the new floorspace was allocated to various centres on the basis of protecting or enlarging upon the traditional hierarchy. This means that a large number of the plans were both mechanistic in their style and pre-occupied with shaping the physical aspects of shopping. Such issues as the extent to which new retailing developments might improve or impair the employment base; whether they would enhance the choice in shopping and achieve greater cost savings for disadvantaged consumers; and how far they could be integrated with the pattern of wholesaling and warehousing to achieve a greater efficiency in distribution, were matters that received scant attention.

The structure plans, of course, are subject to continuous revision and there has recently been much back-tracking on some of the earlier optimistic forecasts made (Shaw and Williams, 1980). The initiative for introducing a new conceptual approach to the business of retail planning, however, has clearly passed to the Districts. So far, we have seen from these authorities mainly a series of studies or topic reports pending the integration of various shopping proposals with those for other land use activities into a formal set of plans. These display much less regard for quantifying the optimum size and arrangement of shopping centres although much attention remains focused on improving their physical characteristics. This is inevitable at this level since such matters as the attractiveness of the environment, the amount of car parking available, the provision of rear-access delivery, and the scope offered for pedestrianisation are important considerations for consumers and retailers alike (British Multiple Retailers Association, 1980). Emerging from these topic reports, however, appears to be a much greater concern for the socio-economic problems surrounding retailing, such as the difficulties experienced by small, independent shops and the lack of access to essential services for people living in isolated communities. What seems to be still missing, nevertheless, are some firm proposals as to how these problems may be overcome.

The clearest signs of a new era of planning emanate from the draft proposals which have recently been put forward for the central areas of several provincial cities (notably Newcastle, Manchester and Leeds). It is within these centres more than anywhere else that there are likely to be severe constraints on future growth, as we have indicated. Gone are the heady ideas of the 1960s for massive redevelopment, the building of vast office complexes as well as new shopping schemes, the segregation of shoppers from traffic through elevated walkways or subterranean malls. In their place, the plans for the 1980s will stress the renovation of outworn properties, the accommodation of new retailing activities on a small, piece-meal basis and the introduction of limited traffic management schemes rather than major road-building programmes (Burns, 1979). There will also be an emphasis on the need to attract new job-generating industries, to conserve the more important architectural legacies of the past, and to promote the central area as an important cultural, leisure and tourist centre (Holliday, forthcoming). While these features are in themselves desirable, however, they represent a general set of aspirations to be fulfilled by the plans and the contributions which the control of retailing can make to them remain unclear.

During the next few years, therefore, we can expect retail planning to be less ambitious than it has been in the past, at least in terms of its prescription as to what new forms of development should be accommodated and where. How long this lower profile will be maintained is difficult to foresee. Looming on the horizon is the threat of massive technological change which, if some soothsayers are correct, will lead to a drastic decline in the office-based workforce and hence central area sales, together with a radical alteration to the structure of retailing in the suburbs induced by the introduction of home-based computerised shopping services.

The Traditional Themes of Geographic Enquiry

Like planning, the literature in retail geography has been dominated to-date by a concern for the physical aspects of retailing activity. There have been three main themes where the two fields of study have overlapped: in descriptive assessments of the general pattern of retailing throughout the city; in analyses of the impact on the traditional environment of new types of development; and in the identification of specific kinds of retailing problems to be found in different parts of the urban area.

The General Pattern of Retailing

Those studies which have been made of the general pattern of retailing have tended to be theoretical in perspective and have been heavily influenced by the work of Berry (for example, 1963) on cities in the United States. They have been addressed primarily to identifying the

principal spatial components of the retail system and examining both
the forms and functions of different configurations. Three broad
types of conformation have been focused upon: a set of nucleated
centres, catering mainly for domestic shopping requirements and
which are located in highly centralised positions to surrounding trade
areas; a set of ribbon activities, comprising a wide range of services as
well as shops, and which are strung out alongside major roads to cater
to passing traffic as well as local residents; and a set of specialised
functional areas, representing like kinds of businesses, which are mainly
located within the central areas. The three types of conformations may
be associated together and described collectively within the context
of a general hierarchy of business complexes, but from the point of
view of central place theory the hierarchy principle is most appropriate
to the nucleated centres. The ribbon developments are less prominent
than their counterparts in North America, are mainly relicts of a past
legacy of transport influences (particularly from tram networks) and
are mainly confined to the inner city. They have been subject to much
clearance and redevelopment during the last two decades, and their
absence in suburban areas has been due to restrictive planning controls.
The specialised functional areas have sometimes been reinforced and in
other cases depleted by the process of redevelopment within town and
city centres.

These features of the retail pattern were first elucidated in a series
of case studies of Coventry (for example, Davies, 1974), since when
three more specific research avenues have been followed. The first
concerns an attempt to apply the mathematical axioms of central place
theory to the general hierarchy of centres in urban areas. Following
the more broadly-based work of Parr (1978), Warnes and Daniels
(1979) have suggested that a combination of Christaller's rules
regarding the frequency and spacing of different sizes of centres may be
suitable for explaining the complicated relationships to be found at the
urban level. In particular, they consider that variable K-ratios of 7 (the
administrative principle), 4 (the transport principle) and 3 (the market
principle) may provide a satisfactory explanation of the number of
centres to be found at second, third and fourth order levels in the
hierarchy respectively. This seems to be intuitively sound on the
grounds that second order centres are mainly district-level centres that
will have been subject to much planning control; third order centres
comprise a mixture of neighbourhood centres and ribbon developments;
and fourth order centres embrace the numerous local parades and
scattered corner stores that have sprung up in response to small pockets

of demand. The second avenue of research has involved an attempt to introduce a more dynamic perspective to the interpretation of the retail pattern. This has been taken furthest by Potter (1982) who has developed a stage model of the evolution of the retail system, drawing on the historical studies of Wild and Shaw (1979). The third avenue of research has involved a refinement of the techniques of analysis to be used in classifying different kinds of business configurations (for example Potter, 1980). In his exhaustive studies of Stockport, it is also significant that Potter arrived at the same conclusions as Warnes and Daniels regarding the applicability of variable K-ratios for explaining the structure of the hierarchy of centres.

Collectively, this body of research has established a comprehensive framework for the study of the spatial characteristics of the retail system but it is a framework that has been used mainly to look backwards to the outcome of a past set of processes rather than forwards to the accommodation of new agents of change. It has been shown, however, to have clear relevance to the spatial design aspects of planning (Davies, 1977) and should be particularly appropriate for guiding the analytical stages of forthcoming district or local plans. In distinguishing between different kinds of business conformation it provides a basis both for identifying a variable set of environmental problems to be found and for recognising the preferred locational settings of various kinds of retailing activity.

The Impact of New Development

The large number of studies which have been made about new types of development, nevertheless, have rarely referred to this classification of the retail system as a whole. While their central focus has been on the impact of new development on the traditional pattern of retailing, their terms of reference have mainly been the general hierarchy of centres. In a planning context, the main aim of the studies has been to see whether a new development can be integrated into the existing hierarchy or to form an extension to it, and if not what would be the likely scale of any disruption caused. Some observers (for example, Dawson, 1979), however, have contended that most new development, particularly superstores and hypermarkets, are so different from the traditional methods of retailing that any attempt to reconcile them to the existing structure of centres is rather meaningless.

The extensive literature that now surrounds the impact of superstores and hypermarkets was originally prompted in the later 1960s by a general debate at that time as to how much decentralisation of the retail trades should be permitted in Britain. To resolve the debate it seemed almost as if central government allowed a few stores to be built so that their actual effects could then be assessed. Amongst these were the Carrefour developments at Caerphilly and Eastleigh which were closely monitored by Lee and Kent (1973–9) and Thorpe and McGoldrick (1974) on the one hand and Wood (1976–8) on the other. Each of these studies came to the conclusion that the hypermarkets had relatively little effect on the surrounding hierarchy; and moreover that most of the adverse trading repercussions were absorbed by the branches of multiple firms based in larger centres rather than the independent shops found in smaller centres. Subsequently, as more developments were opened, further studies were undertaken using a variety of techniques to see whether the scale of effects might vary for different sizes of stores in different types of locational setting. Somewhat surprisingly, the bulk of the findings have been similar to those reported earlier (Seale, 1978; Schiller, 1981); although it must be recognised that those developments given planning approval were anticipated as having less serious trading repercussions than those that were refused.

Local authority planning departments have formed widely differing attitudes, however, towards superstores and hypermarkets (Davies and Sumner, 1978). Despite some 40 impact studies which have been made, there is still considerable uncertainty as to what the relative benefits versus the disadvantages of these developments have been. The impact studies appear to have become too parochial in their concern simply for examining a set of trading effects. A broader-based enquiry is required that takes stock of the way in which these developments have enhanced or reduced consumer opportunities within any community; how far they have assisted in or stifled the generation of alternative new retailing activities; and what sorts of changes they have brought to the structure of retail employment.

A similar assessment would be desirable with respect to the impact of purpose-built, managed shopping centres. This is a topic, however, about which much less has been written. About 15 years ago, there was much controversy over the possible introduction into Britain of 'out-of-town' shopping centres of the North American kind; but the threat which these posed to the relative trading health of the central area and the visions conjured up over their promotion of urban sprawl, led to a fierce and united planning opposition to them which has continued through to the present day. To meet the pressure for new development within the durable-goods trade sector, however, the local authorities encouraged the construction of new centres within central areas (or alternatively as district centres in residential areas) as has been indicated. Quite remarkably, there has since been but one in-depth study of the impact caused by the town-centre schemes. This was an investigation of the local and regional trading effects of the Eldon Square Shopping Centre in Newcastle upon Tyne (Bennison and Davies, 1980). This found that although the centre altered patterns of expenditure quite radically in its first year of opening, during the next three years there was a substantial recovery amongst those shops and surrounding shopping centres that were most adversely affected. Eldon Square appeared to enhance the overall attractiveness of the central area, thus drawing in additional custom; and further afield, those shopping centres which bore the brunt of competition were primarily those in market towns or commuting settlements which were experiencing considerable population growth, and hence could absorb the effects relatively easily.

Given that relatively fewer town centre shopping schemes will be built in future, it may seem unnecessary or too late to embark on further studies of their effects. These could be important, however, in two main respects. First, a number of small developments will continue to be constructed in smaller urban settings and the impact of these on the traditional environment might be proportionately greater than in the Eldon Square case (Schiller, 1975). Secondly, and more fundamentally, one will still want to know with respect to many schemes already completed what their inhibiting effects will be on the growth potential for retailing activities in other parts of the urban area. There is the danger that with so much investment now having gone into town-centre schemes that the main thrust of future planning policies will be aimed at protecting these and resisting all forms of complementary activities that seek to become established elsewhere.

Specific Problems in Specific Areas

The relative health of the central area of the largest cities, however, will become increasingly threatened in the future as we have indicated. We can expect to see not only a curtailment of new investment but also a shrinkage in the trade area support for existing shops, stemming in part from the continuing outward exodus of the population and, in the longer term, in part from the contraction of the central area workforce. The implications of this are that the effective action space for shopping will likely become much reduced and the decaying margins of the central area will likely become more extensive. There is the prospect therefore of large tracts of outmoded retailing land uses emerging which, in a climate when redevelopment will be less favoured than renovation, will not be easily replaced. It is difficult to conceive of a new set of leisure, tourist and cultural facilities occupying much of this space although this is probably what will be required in contemplating a wider socio-economic role for the future city centre.

An extensive decaying retailing environment is already to be found in many inner city localities, of course. This is the result of an accumulation of problems, involving the diminution of the resident population, the growth in the proportion of elderly and poorer consumers, a high incidence of vandalism, a surfeit of small shops, ageing building premises, and in some cases planning blight (URPI, 1980c). The worst examples are to be seen in the older, ribbon developments emanating from the central area, but some forbidding conditions have also arisen in many failed shopping centres established as part of comprehensive redevelopment programmes (Benwell Community Project, 1979). When first opened, these new centres were an attempt both to improve the physical setting for retailing and to provide a modern set of shopping opportunities for consumers; but they proved difficult to let either to local independent businessmen or to the multiple groups because of a basic lack of confidence that enough trade could be generated. A recent study (NEDO, 1982) has suggested that a greater trade potential would be realised if superstores could be attracted into such schemes as the key tenant; but it is difficult to see the superstore operators investing in the inner city given the likely future trends for this area that we have already described.

Deteriorating retailing environments and declining shopping opportunities are also becoming more prominent in certain outlying parts of the urban area. It is not possible to be precise about these for no studies have been made of the changes which seem to be occurring; but they are related again to the underlying market conditions to be found. In some cases, the simple process of the ageing of a population has gradually eroded the trading base of a local parade of shops; but in other cases, the appearance of much better shopping opportunities elsewhere has siphoned off the wealthier, car-owning consumers, leaving the smaller traders with a customer support not unlike that found in the inner city. Given the evidence presented above regarding the impact of superstores, however, the effects of the competing attractions may not be as straightforward as this. Many small independent retailers may remain quite viable, but the demise and possible closure of a multiple's supermarket will set in train the first signs of decay.

As far as new small-scale development is concerned, one can recognise the same lack of investment confidence in the building of outlying local parades and neighbourhood shopping centres as is to be found in the inner city (URPI, 1980d). Yet there remains considerable scope for the introduction of different kinds of development in alternative types of locations. The so-called modern convenience stores, mini-markets, fast-food outlets and many car-oriented services might well flourish alongside main roads leading to these areas. This might lead to the emergence of small ribbons of activity, which have hitherto been resisted by planners, but development control measures could be invoked to ensure that these do not become unsightly or a hazard to safety.

Relating Past Practice to Future Needs

The retail system has clearly become more fragmented in recent years with strong contrasts appearing particularly between a declining set of older, smaller activities and the growth points of new, large developments. The socio-economic problems that stem from this will become an increasingly dominant focus of geographical enquiry in the future. The central issue will be the time-honoured one of how to improve the equity of a planned provision of shopping resources whilst at the same time promoting the efficiency of the process of distribution; but we can also expect to see a response to the new concerns of the 1980s that permeate other fields. The most pressing of these at the present time are the changes taking place within employment and later the full implications of modern technology in altering patterns of consumer behaviour are likely to become apparent.

Equity Versus Efficiency

The issue of equity versus efficiency was a major consideration of those debates in the late 1960s and early 1970s as to whether or not hypermarkets and 'out-of-town' shopping centres should be allowed to be developed in Britain. Hillman (1973) and others warned against the benefits of the new developments mainly accruing to the affluent sections of the population while the weaker sections of society would be left with a declining set of traditional shopping activities. It can be argued too that the whole ethos behind the continuing use of the hierarchy principle to arrange new retail developments in space is because of a belief amongst planners that this arrangement best serves the interests of both equity and efficiency (NEDO, 1971).

Given the erosion of the hierarchy of centres, however, the existence of large numbers of hypermarkets and superstores, and their likely continued growth in the future, the issue has become more complicated than it was previously. Rather than restricting new developments, they may now have to be used as the instruments by which equity and efficiency are enhanced. A distinction needs to be made, nevertheless, between the problems that have accrued to the various parties involved, namely the consumers, retailers and the planners themselves.

From the consumer's point of view, the issue has become one of a basic inequality in the access afforded different population groups to the lower prices and greater choice in goods provided by hypermarkets, superstores and other large scale activities. Significant numbers of people may now be deemed to be disadvantaged consumers by virtue of their lack of access to these facilities (Davies and Champion, 1980). The majority are to be found in the inner city where there is little prospect as we have indicated of new developments being established in the future. Two possibilities present themselves. One is to encourage the opening of new developments on the fringe of the central area or on the outer margins of the inner city where a larger catchment population could be served. The other is to introduce special services for more housebound consumers, involving either the use of conventional transport forms or the use of the telephone network to effect an ordering and delivery system. An experimental service has been created in Gateshead based around the use of micro-computers and telecommunication facilities, whereby relatively immobile consumers can order their goods from a distant superstore through local branch libraries (Gateshead MBC, 1982).

Amongst the business community, one can also identify a growing number of disadvantaged retailers in the sense of those small, independent shopkeepers that lack access to the larger-scales of economy that affiliated stores or multiple groups enjoy. It may not be possible to modify the physical environment to a sufficient degree to alleviate their problems but some relaxation of development controls along the lines already indicated may help. Some extensive studies which have been made of their problems, however, suggest that their destiny is rather within their own hands, by virtue of their need to improve their trading skills and to eliminate unnecessary competition (Dawson and Kirby, 1979). Indeed, Kirby (1980) has argued that there is likely to be an increased scope for new small business formation in the future as more and more large stores become built. This stems from the polarisation theory which assumes that because the large stores will only be intermittently spaced a substantial trade vacuum will emerge at the local level. As evidence of this, he cites the experience of the United States where the so-called modern convenience store has been the fastest growing of all retail developments in recent years (Kirby, 1976).

For the planner, faced with the need to foster efficiency amongst both large and small businesses and to ensure that equitability in trading opportunities is maintained, the problems are mainly those of identifying appropriate policies and procedures through which these goals can be obtained. There are two requirements to attend to. First, the weakening of the hierarchy at both its upper and lower levels, together with the relaxation of development controls, means that a revised set of spatial design concepts will have to be formulated. Ideally, this would involve reducing the prominence given to the hierarchical principle and devoting more attention to the different locational characteristics of new firms. Secondly, there must be some greater regard for the aspatial aspects of retailing, revolving around its contribution to the improvement of people's living standards both from the point of view of consumer savings on the purchase of goods and through the generation of jobs and the creation of additional purchasing power. A number of geographical studies in the past have looked at the price variations of goods in different sectors of retailing (for example, Parker, 1974); to-date there has been much less concern for the employment opportunities created by the industry.

Changes in Retail Employment

There appear to be two areas of research which are currently being pursued with respect to employment in retailing. Given the absence of any foundation work on this topic, the first revolves around an exploration of what trends have been taking place within the industry as a whole, in terms of both the total number of employees to be found and the structural characteristics of the workforce (Davies and Wade, forthcoming). The main indications are that the mid-1970s once again proved to be a watershed in so far as a change in the direction of these trends is concerned. Prior to this time, there was a massive growth in female part-time labour and consequently a dramatic increase in the total number of jobs, linked primarily to the changes in shopping behaviour that had occurred but also aided by the recruitment policies of expanding multiple firms. Since then, however, the rate of growth in female part-time labour has tapered off and during the recession the total number of jobs has become reduced. More ominously for the future, fewer younger people have been employed than in the earlier period with most firms now showing a marked preference for more mature adults, whether recruited on a part-time or full-time basis.

The second area of research involves a more specific assessment of what the employment contributions and the employment effects have been of new retailing developments, most notably hypermarkets and superstores. This constitutes in part an outgrowth of the impact studies reviewed earlier but also a response to what changes are likely to occur in retail employment more generally in the future given the continued growth of these developments. Three sets of findings have emerged so far but their results are not conclusive and need to be examined in different kinds of case-study settings. The first is that whilst these large developments are dominated by part-time staff, the proportions of part-timers to full-timers are not significantly different from those in smaller sizes of stores, nor from those in conventional variety stores, and the incidence of males may be higher. Secondly, despite the claims made in some impact studies (for example, Brent Community Law Centre, 1978) that the new developments will ultimately lead to a net loss of jobs in local economies due to their trading repercussions, most evidence points to their providing a net gain. Thirdly, there is some indication that the developments contribute to a raising of pay and service conditions and lead in an indirect way to an improvement in skills at the management level.

There are clearly many other avenues surrounding retail employment that need investigation in the future. Among these are the changes which have occurred in different sectors of the industry and in different areas of the city. In the longer term, the introduction of new technological innovations both within and outside of retailing might have quite catastrophic repercussions on the work practices of employees and there could be a marked shrinkage of the labour force if the trading health of city centres becomes seriously undermined.

The Spectre of Technological Change

The references that have so far been made within this essay to the advent of modern technology suggests that this holds much promise with respect to improving people's accessibility to new retail developments but that at the same time it poses a threat to those who work in more traditional kinds of retailing activities. These are essentially speculative observations, however, and several years will need to pass before the truth unfolds. One must not minimise the operational problems that are involved in establishing remote forms of shopping, nor the resistance to change that members of the workforce can bring.

Tracing through those technological innovations that have so far been introduced into retailing, nevertheless, rapid progress has been made in computerising warehousing and stock control procedures and the industry is poised to improve its selling operations through the installation of EPOS (electronic point of sale) equipment (NEDO, 1982). The next logical step, given the investments that are now being made into cable TV, telecommunications systems and micro-computer facilities, is for the ordering side of the shopping function to become automated. There are several ingredients that will help this along: the increased use of credit cards for payment of shopping goods, the familiarisation of the public with automated ordering devices in the travel industry, the growth in telephone ordering of mail-order goods advertised on radio or television (Kirby, 1982). It is not yet clear as to which types of shopping activity the development of computerised ordering facilities will be most appropriately applied, but the success of catalogue showrooms may be taken as one guide and the routine purchasing of grocery goods as another.

The technical challenge of introducing computerised ordering facilities into people's homes, however, is small in comparison to the likely difficulties to be faced at the supply end. The demand for a limited range of high value goods may be relatively easily met, but meeting an assortment of orders for grocery goods would be an entirely different proposition. One would have to conceive of a warehouse operation that was itself fully computerised and where the packing of goods could be achieved mechanically. Then there are the problems of assembling packages of goods at collection points or organising them into batches to be delivered to people's homes.

If and when many of our current routine shopping activities can be conducted from the home, however, both the geography and planning of the retailing environment will become substantially altered. It will be salutary now to start thinking in depth as to what these alterations are likely to be. Looking back over the last decade, it seems as if we were caught unprepared for the changes that have emerged in retailing in just the last few years and we ought not to repeat this mistake.

Conclusion

This essay has described a new set of circumstances surrounding the structural characteristics of the retail system in British cities. These emanate primarily from the demographic and economic changes of the second half of the 1970s, the differential growth of new types of retailing developments, and what we might refer to as a retreat from past forms of planning intervention. The general result has been the emergence of a more heterogeneous physical environment than has hitherto been seen and a more glaring set of socio-economic problems that will need to be resolved. The traditional themes of geographical research appear to have provided a firm foundation from which the changes in the physical environment can be monitored and explained; but we can be less certain of our ability to diagnose the new found social and economic ills and prescribe an appropriate remedy for them. At the same time, we lack for the moment a detailed knowledge and understanding of the changes taking place in retail employment; and the current love affair with the potential impact of technological change looks to be more flirtatious than the outcome of serious thought.

References

Bennison, D.J. and Davies, R.L. (1980) 'The Impact of Town Centre Shopping Schemes in Britain: Their Impact on Traditional Retail Environments', *Progress in Planning, 14*, 1–104

Benwell Community Project (1979) *From Blacksmith's to White Elephants: Benwell's Changing Shops*, Benwell Community Project, Newcastle-upon-Tyne

Berry, B.J.L. (1963) *Commercial Structure and Commercial Blight*, Research Paper 85, University of Chicago, Department of Geography

Brent Community Law Centre (1978) *Hypermarket at Neasden?*, Brent

British Multiple Retailers Association (1980) 'Guidelines for Shopping', BMRA, London

Burns, W. (1979) in *Quality in Urban Planning and Design* (ed.) R. Cresswell, Newnes-Butterworth, London

Davies, K. and Sumner, J. (1978) 'Hypermarkets and Superstores: What do the Planning Authorities Really Think?', *Retail and Distribution Management, 6 (4)*, 8–15

Davies, R.L. (1974) 'Nucleated and Ribbon Components of the Urban Retail System in Britain', *Town Planning Review, 45 (1)*, 91–111

Davies, R.L. (1977) 'A Framework for Commercial Planning Policies', *Town Planning Review, 48 (1)*, 42–58

Davies, R.L. and Bennison, D.J. (1979) *British Town Centre Shopping Schemes: A Statistical Digest*, Unit for Retail Planning Information, Reading

Davies, R.L. and Champion, A.G. (1980) *Social Inequities in Shopping Opportunities: How the Private Sector Can Respond*, Tesco Stores Holdings Ltd., Cheshunt

Davies, R.L. and Wade, B.F. (forthcoming) *Retail Employment Change in Scotland*, Scottish Office, Edinburgh

Dawson, J.A. (1979) *The Marketing Environment*, Croom Helm, London

Dawson, J.A. and Kirby, D.A. (1979) *Small Scale Retailing in the UK*, Saxon House, Farnborough

Gateshead M.B.C. (1982) *Shopping and Information Service*, Gateshead

Guy, C.M. (1980) *Retail Location and Retail Planning in Britain*, Gower, Farnborough

Henley Centre for Forecasting (in preparation) *Planning for Consumer Markets*, Henley

Hillier, Parker (1979 and 1981) *British Shopping Developments*, Hillier, Parker, London

Hillman, M. (1973) 'The Social Costs of Hypermarket Developments', *Built Environment*, *2*, 89-91

Holliday, J.C. (forthcoming) 'City Centre Plans in the 1980's' in *The Future for the City Centre* (ed.) R.L. Davies and A.G. Champion, Academic Press, London

Kirby, D.A. (1976) 'The Convenience Store Phenomenom', *Retail and Distribution Management*, *4 (3)*, 31-3

Kirby, D.A. (1980) 'The Future of Small Unit Retailing in the UK: the Implications for Planning' in *Local Shopping Centres and Convenience Stores: Report of an URPI workshop*, URPI, Reading

Kirby, D.A. (1982) 'Shopping and the Micro Chip', *Town and Country Planning*, *51 (1)*, 10-13

Lee, M. (forthcoming) 'Property Development in the 1980's' in *The Future for the City Centre* (ed.) R.L. Davies and A.G. Champion, Academic Press, London

Lee, M. and Kent, E. (1973-9) *Caerphilly Hypermarket Study*, Donaldson, London

National Economic Development Office, Distributive Trades EDC (1971) *The Future Pattern of Shopping*, HMSO, London

National Economic Development Office, Distributive Trades EDC (1981) *Retailing in Inner Cities*, NEDO, London

National Economic Development Office, Distributive Trades EDC (1982) *Technology: The Issues for the Distributive Trades*, NEDO, London

Pacione, M. (1979) 'The In-Town Hypermarket: An Innovation in the Geography of Retailing', *Regional Studies*, *13*, 15-24

Parker, A.J. (1974) 'Intra-Urban Variations in Retail Grocery Prices', *Economic and Social Review*, *5*, 393-403

Parr, J.B. (1978) 'Models of the Central Place System: A More General Approach', *Urban Studies*, *15*, 33-50

Potter, R.B. (1980) 'Spatial and Structural Variations in the Quality Characteristics of Intra-Urban Retailing Centres', *Transactions of the Institute of British Geographers*, *4 (2)*, 207-228

Potter, R.B. (forthcoming) *The Urban Retailing System: Location, Cognition and Behaviour*, Gower and Retailing and Planning Associates, Aldershot

Randolph, W. and Robert, S. (1981) 'Population Redistribution in Great Britain, 1971-1981', *Town and Country Planning*, *50 (9)*, 227-231

Schiller, R. (1975) 'The Impact of New Shopping Schemes on Shops in Historic Streets', *The Planner*, *61*, 367-9

Schiller, R. (1981) 'Superstore Impact', *The Planner*, *67*, 38-40

Seale, S. (1978) *The Impact of Large Retail Outlets on Patterns of Retailing: A Synthesis of Research Results in Great Britain*, Scottish Office, Edinburgh

Shaw, G. and Williams, A. (1980) 'The Structure Plan Offers More Scope for a Sane Approach to Retail Planning', *Retail and Distribution Management*, Jan/Feb., 43-7

Thorpe, D. and McGoldrick, P.J. (1974) *Carrefour: Caerphilly – Consumer Response to a Hypermarket*, Research Report 12, Manchester Business School, Retail Outlets Research Unit

Unit for Retail Planning Information (1980a) *Consumer Retail Expenditure Trends*, Information Brief 80/9, URPI, Reading

Unity for Retail Planning Information (1980b) *List of Hypermarkets and Superstores*, URPI, Reading

Unit for Retail Planning Information (1980c) *Retailing in the Inner Cities: Report of Two URPI Workshops*, URPI, Reading

Unit for Retail Planning Information (1980d) *Local Shopping Centres and Convenience Stores: Report of an URPI Workshop*, URPI, Reading

Warnes, A.M. and Daniels, P.W. (1979) 'Spatial Aspects of an Intra-Metropolitan Central Place Hierarchy' in *Progress in Human Geography*, *3 (3)*, 384–406

Wild, M.T. and Shaw, G. (1979) 'Trends in Urban Retailing: the British Experience During the Nineteenth Century', *Tijdschrift voor Economische en Sociale Geografie*, *70 (1)*, 35–44

Wood, D. (1976–8) *The Eastleigh Carrefour: A Hypermarket and its Effects*, Department of the Environment, Research Reports 16 and 27, London

7 TRANSPORT

P.R. White

Introduction

In recent years a much more comprehensive approach to the study of urban transport has been adopted. Previously, both geographers and transport planners had tended to concentrate only on motorised travel, especially in any statistical work, but as the viewpoint has gradually shifted to that of a fuller understanding of traveller behaviour, the importance of walking and cycling — even in the largest cities — has been acknowledged. Similarly, journey-to-work demand had been seen as crucial, both in transport planning (creating the peak demand for infrastructure) and among many geographers, for example in studies of household location and urban growth as transport networks grew. Although the peak journey-to-work demand remains significant, it is important to recognise that it represents only a minority of all travel. Access to schools, shops and entertainment is determined by the price and quality of transport services, and also affects household location decisions. Within 'western' cities an ageing population, and perhaps long-term changes in the nature of full-time employment, reduce the relative importance of the work trip still further. Within 'developing' countries urban growth continues at a high rate (Pacione, 1981), increasing demands for all types of trip.

This review begins by examining trip length distribution, and from this, the importance of different modes, and effects of city size. Differences in travel behaviour according to person type and house-hold structure are then examined, together with change over time in such patterns. Shifting from the demand to supply side, the review then looks at finance and system management. Finally, changes in technology and possible future trip patterns are discussed.

The Demand for Transport

Trip Length Distribution

A good indication of the overall distribution is given in the National Travel Survey (NTS), conducted on four occasions since 1965/6 in

Britain. This is a comprehensive record of trips from a large sample of households throughout the country, from which many inter-relationships can be derived and examined. The most recent for which data is available is that of 1978/9, a valuable feature of which was that walking trips were included more thoroughly than in many earlier studies, which tended to omit walking and cycling entirely, or include them only for longer work trips. The data illustrate very clearly the rapid fall in trip rate as distance increases, some 52 per cent of all trips being under 2 miles (3.2 km) in length. Although the NTS sample also includes rural areas, it is predominantly urban in character, and the picture shown may be considered representative of urban travel as a whole.

The fall in trip rate as distance rises is approximately proportional to the inverse square of distance, as would be expected from the simple formulation of the gravity model. The omission of short walk trips from many other surveys can present a misleading trip length distribution, in which the peak trip rate (being that for motorised trips only) is found in the range of one to two miles. This leads to errors in model specification, and undue complexity therein (Goodwin, 1980). If only motorised travel is considered, then such a distribution would of course be logical. Even in the West Midlands, a large conurbation of 2.7 million inhabitants, the average bus trip is quite short — a mean of 4.4 km, and mode of about 2 km (1980-1) in the very skewed distribution. Rail trips in the West Midlands have a greater mean length of 10 km, but account for only 4 per cent of all public transport trips.

These distributions serve to emphasise the essentially 'local' nature of urban travel, even in an age of high car ownership and good public transport availability. In the 1978/9 NTS walk trips were found to comprise no less than 39 per cent of the total (this itself may be an understatement), being of particular importance in shopping (46 per cent) and education trips (61 per cent), purposes which also had above average proportions below one mile in length (40 per cent and 51 per cent respectively). Conversely, only 17 per cent of work trips were under one mile long (Transport Statistics Great Britain 1970-80).

The Role of Motorised Modes

Although walking and cycling now receive greater emphasis, especially in the proportion of total travel time they occupy (about one third), they are considerably less important as a proportion of the resources

Table 7.1: Trips by Motorised Modes in Britain (Percentage by Mode)

Trip category	Bus	Rail	Private car, etc.
NTS 1978/9 all purposes	22	2a	74
NTS 1978/9 Shopping/ personal business	25	2a	75
NTS 1978/9 Education	51	3a	46
NTS 1978/9 Work	24	6	69
NTS 1975/6 (by size of area) Work (with education in brackets)			
London	24 (56)	22 (6)	53 (36)
Over 250,000	32 (64)	1 (4)a	66 (34)
100,000 to 250,000	26 (67)	4 (–)a	69 (35)
25,000 to 100,000	19 (57)	5 (5)a	76 (39)
3,000 to 25,000	17 (54)	3 (3)a	79 (43)
Rural	13 (67)	1 (0)a	86 (33)
All areas	23 (60)	5 (3)a	71 (38)
NTS 1975/6 (all purposes, by type of area)			
London built-up area	21	11	68
Birmingham built-up area	26	2	72
Manchester built-up area	31	2	67
Glasgow built-up area	41	5	54
Liverpool built-up area	37	2	61
Urban 250,000 to 1 million	27	2	71
Urban 100,000 to 250,000	24	2	74
Urban 50,000 to 100,000	18	3	79
Urban 25,000 to 50,000	18	2	80
Urban 3,000 to 25,000	13	2	85
Total Urban	21	3	76
Rural (up to 3,000)	11	2	87
West Yorks. study 1975 (all purposes)	38	3	51
Mid-1960s studies (all purposes)			
Kingston-upon-Hull	37	—	63
Cardiff	37	1	62
Norwich	29	—	71
Cambridge	23	—	77
Plymouth	53	—	47
Coventry	39	—	61
Brighton	49	—	51
Edinburgh	51		49
Merseyside	(54)		46
Glasgow	(72)		28
London	(60)		40

Note: a. Very small proportion — estimate may be unreliable.

Table 7.2: Internal Urban Public Transport Trips, Split by Purpose (Percentage)

Trip category	Work	Education	Shopping/personal business	Other
NTS 1978/9 al. areas	28	12	32	28
Oxford & county 1977	37	19	26	18
S.W. Glasgow suburbs 1980 (Scotmap)	46	11	15	28
West Yorks. 1975	41	13	16	30
Mid 1960s				
Kingston upon Hull	36	15	24	25
Exeter	28	NA	35	37
Cambridge	24	19	24	33
Norwich	40	10	23	26
Northampton	34	38		
Brighton	38	14	26	22
Coventry	42	24	17	17
Plymouth	40	13	20	27
Leicester	38	NA	25	37
Belfast	45	7	16	32
London 1971				
Underground	72	4	9	15
Bus	53	9	22	18

devoted to transport (as distinct from benefits obtained). For purposes of comparison with earlier studies in specific cities, and with studies outside Britain, it is desirable to analyse motorised travel only, since the non-motorised modes are often omitted altogether, or treated in an inconsistent fashion. Table 7.1 shows the proportions of all trip purposes, and selected purposes, by bus, rail and private motor vehicles (mostly the car). Aggregate results from 1975/6 and 1978/9 NTS data are compared with selected British urban studies.

Within this sector, private vehicles are generally predominant but public transport (bus, plus rail and rapid transit in larger cities) continues to play a major role. In many studies this has been linked very closely with the journey to work, but as can be seen, its greatest share in 1978/9 was for education trips (mostly by older schoolchildren, to schools beyond walking and cycling distance). The share of work trips was 36 per cent and of shopping/personal business trips, 28 per cent. However, as Table 7.2 (showing the split by purpose within the public transport market) indicates, shopping and personal business trips account for a greater share of all public transport trips (32 per cent) than work (28 per cent), since the total number of shopping and personal business trips (by all modes) is substantially greater than that of work trips.

Table 7.1 also shows results from the 1975/6 NTS, which have been analysed in more detail (National Travel Survey, 1979). There does not appear to have been any significant change in public transport's share of the work trip market (over a longer period, this has been falling), although for education it has fallen, possibly as a result of more parents giving lifts to and from school.

The relative importance of shopping and personal business trips vis à vis work trips within the public transport market may seem surprising, given the radial form of most public transport networks and peak journey to work demand, with limits on use of the car. As shown below, the traditional stereotype is true of the larger cities, but in many small to medium sized centres shopping/personal business and work trips account for a similar share of total daily demand. The picture is partly explained by car availability. In Britain in 1980, 42 per cent of households had no car, 44 per cent one car, and 13 per cent more than one. The typical household may thus be considered as a one car household in which the car is either used for the peak work trip, being unavailable to other members during the working day, or left at home for trips such as shopping. As the head of household often has first claim on the car, it tends to be used for the work trip, except where restrained, or public transport becomes more attractive for this purpose (usually true in the larger cities). Many shopping and school trips thus become semi-captive to public transport, if over distances above walking length. A complementary pattern of car and public transport use is thus found. At evenings and weekends, car availability is less restricted, all or most members of the household tending to travel together for social visits, etc.

The pattern is not a completely rigid one. Schoolchildren may be given lifts to school when their journey coincides with the adults' journey to work, returning in the afternoon by bus. Car-sharing may affect the pattern, enabling some cars to be left at home during the working day for use in shopping, and likewise such trips themselves may be made by car-sharers. Where there is more than one person in the household making a peak work or education trip, then one may be by public transport, the other by car. Alternatively, the household may possess two cars, one quite possibly being provided by an employer for 'business' use, both being used at the peak.

Table 7.1 also shows the role of bus and rail within the public transport sector. In Britain, rail accounts for a very small proportion of all trips at national level. Most of these are concentrated in the London and South East Region (550 million trips per annum on the LT underground, and about 500 of the 700 million per annum on British Rail), and within London, as the breakdown of the 1975/6 NTS shows, rail and bus account for very similar market shares. In other areas, the very small rail sample leads to unreliable estimates, although a significant role is played in Merseyside, Glasgow and Tyne and Wear. Note that some other European countries display a very different pattern, with rail taking a large share of public transport trips in cities of about one million upward, especially in West Germany and Scandinavia.

The public transport share tends to decline as size of urban area falls, from almost 50 per cent of all work trips in the London area, to 20 per cent in towns between 3,000 and 25,000, and 13 per cent in rural areas. However, in the size range 25,000 to over 250,000 the range is relatively small (20 to 33 per cent), and some of the denser, more nucleated centres of around 100,000 can have almost as high a public transport share as the conurbations.

Examples of specific cities in Table 7.1 show public transport's share at 50 per cent of all trips in the West Midlands, and (in 1975), 38 per cent in West Yorkshire. Earlier studies from the mid 1960s tend to exaggerate the present market share of public transport, but nonetheless, it is interesting to see that centres such as Brighton and Plymouth (about 200,000 population) attained shares almost as high as the Merseyside conurbation.

Travel Behaviour by Person and Household Type

A characteristic of many transport studies of the 1960s and early 1970s was that demand was treated only in aggregate terms, with little understanding of journey purpose or type of traveller. Some subdivision was attempted, but often on crude lines. For example, the typical 'category analysis' approach classified households by three components — number of members (and number thereof in employment), income level and car ownership. Although fairly strong statistical relationships could be found — for example, an increasing number of trips per day as household size grew, and as income rose — the understanding of individual patterns was very limited. Thus a two-person household without a car could be a pair of students, or pensioners, with markedly different demands for travel. Distribution of trips by time of day varies markedly according to journey purpose and type of traveller, with consequent effects on the transport network. An

Table 7.3: 'Lifecycle' Classification of Households for Transport Study Purposes

Lifecycle group	Description of group	Definition of group
A	Young adults, no children	Head of household under 35
B	Adults with pre-school children only	Child(ren) aged between 0 and 4 years
C	Adults with pre-school and older children	Youngest child between 0 and 4 years, other child 5 or over
D	Adults with primary school children	Youngest child aged between 5 and 11 years
E	Adults with secondary school age children	Youngest child aged between 12 and 16 years
F	Families of adults, all of working age	Youngest person over 16 years
G	Adults, without children of working age	Head of household over 35 years, but less than 65 (male) or 60 (female)
H	Elderly persons	Head of household over 60 (female), 65 (male)

Source: Jones, Dix, Clarke and Heggie (1980).

increase in retired people creates more 'off peak' travel, whereas educational trips follow a more peaked pattern, like those to work.

Within households, the pattern of car availability as discussed above, affects modal choice and trip timing. A particular constraint faced by many households in planning their trips is the presence of children, especially those needing to be taken to/from school. This in turn constrains the timing of other trips, by one or more adult members. A classification into 'life cycle stages' has proved very useful in illustrating the different household structures and their effect on demand. Table 7.3 shows the classification system developed by the Oxford Transport Studies Unit.

By use of detailed household interviews, it is possible to illustrate typical activity/travel patterns during the working/school day for such households, and how decisions of the members' interact. For example, given the need to get a child to/from school, the morning work trip by car may be timed and routed to give the child a lift. In the afternoon, its mother's activities are constrained by the need to walk to fetch the child home.

Very large scale surveys of the type traditionally used in urban transport studies are not practicable if this degree of detail is to be applied to household interaction. However, even a small sample may lead to useful insights — for example, how households react to changes in car availability, or retiming of the school day. A similar approach has been taken in recent studies in California, with 'typical' households or individuals being defined, and their reactions to change studied as a guide to the aggregate change in response to policy changes.

In general, it can be seen that for intra-urban movement, the timing of activities poses the main constraints, with frequency, speed and availability of the different transport modes determining their use, rather than price.

It should be stressed at this point that both the traditional transport studies (focused upon peak infrastructure demand), and more recent household studies, have tended to concentrate on the typical 'weekday' (Monday to Friday) trip pattern, since this contains the peak demand, and also the greatest constraints. However, from the viewpoint of a public transport operator, for example, weekend traffic levels may be important. Here, demand is much more responsive to price changes, with a high proportion of 'optional' trips.

In the long run, trip patterns are much more flexible and responsive to changes in modal characteristics than surveys based only on present patterns may suggest. Households change in structure, passing through the lifecycle stages identified in Table 7.3, and relocate, often with simultaneous changes in work or school destinations. Even in a zone of apparently stable land use, such as a well-established residential area, constant change is occuring as individual households relocate. On a typical urban bus or rail service, as many as 20 per cent of the peak users may have begun to use the service for the first time within only the last twelve months (Maltby *et al.*, 1978; Nash *et al.*, 1981). This shift is not the result of changes in mode in most cases, but a change in job, or household location.

From the viewpoint of the policy maker who may be trying, for example, to encourage a shift between modes, it may be that in the very short run households and their activities are largely fixed, with little inclination to change mode for a journey already established as a routine. However, where other changes occur, leading to changes in route and/or mode used, then a greater opportunity exists to influence mode choice. Over a period of several years, a substantial proportion of users may be influenced in this way, as changes are made. This in turn implies that policies pursued consistently over a number of years — for example, in setting a high standard of public transport service — are likely to have much greater effects than changes introduced only for short 'experimental' periods. It is particularly unfortunate in the

British, and to some extent, American, case that major policy changes have often been pursued for periods as short as a few months (for example, six months in the case of low fares on London Transport introduced by the Greater London Council in October 1981), making any assessment of their possible long-term effect almost impossible. Conversely, policies elsewhere in Europe have shown much greater consistency, despite frequent changes in ruling party, producing cumulative effects such as high use of public transport stemming from high investment and relatively low fare levels.

Returning to the question of trip rates defined by individual and household type, one should also note that, although detailed surveys are needed to get the full picture, some routine data is also produced in a form permitting partial disaggregation by person type. For example, most public transport operators estimate separately the trips made by children, and by pensioners, due to the different fares charged. Dividing these by the respective catchment populations, annual trip rates for each may be estimated. That for pensioners on free or reduced-fare passes shows a remarkable stability in British urban areas, at about 4 unlinked[a] trips per week. A similar stability may be found in the trip rate by holders of passes or travelcards giving unlimited use of the network for a given period (usually held by those primarily making work trips), at 15 to 20 unlinked trips per week.

The overall average per capita trip rate by public transport is thus a function of the average rates within different groups, and the relative sizes of those groups. The pensioner total is clearly influenced by the total in lifecycle group H, and the child total (being mainly trips to/from school) by those in groups C, D and E. The proportion using public transport for the work trip is a function of numbers in groups A to G, and that share of the work trip market secured by public transport.

Taking the West Midlands for example (the metropolitan county centered on Birmingham), an average public transport unlinked trip rate of 207 was observed in 1978-9. The average for pensioners, some 16 per cent of the total population, was close to this, at 231. Holders

Note: a. The term 'unlinked' is used to describe the manner in which most public transport operators assess use, that is each ride on a vehicle is deemed to mark the start of a new trip (except within some rail networks, where through ticketing is provided). Thus someone using two bus routes to get from home to work, is deemed to have made two trips. The effect is to overstate somewhat the 'linked' trip rate (i.e. that in which trips may consist of one or more rides), especially in larger cities where more interchange occurs (White, 1981b).

of travelcards made about 850 trips per year, but comprised only about 7 per cent of the population (about 15 per cent of the working population), a much lower rate among other working-age adults giving an overall average for this age category of 223. Finally, the average public transport trip rate by those aged under 16 was 94 (White, 1981a).

Effects of City Size

Public transport tends to take a larger market share of motorised travel as city size grows, although within the population range 25,000 to 250,000 the variation can be relatively small. This growing share is a result of several factors:

(1) Lower car ownership rates in larger cities. Cause and effect are mixed here — the denser public transport network itself reduces the need to own a car. At any given income level, the rate is noticeably lower than in rural areas.

(2) The greater distances, especially between home and workplace, reduce the extent to which walking and cycling can be used, increasing the demand for all motorised modes.

(3) As city size increases, not only does the public transport network become larger, but qualitative differences emerge. On trunk routes, very high frequencies are found, and demand sufficient to justify specialised facilities such as underground railways. A higher level of evening and Sunday service can be justified, together with some all-night operations.

(4) Larger cities may also tend to have higher economic activity rates (percentage of population in paid employment).

Table 7.2 shows the internal divisions of the public transport market, split by purpose. The share of work trips is typically around 35 per cent, but is noticeably lower in some smaller centres while rising to about 50 per cent in larger cities. In London, about 53 per cent of bus trips, and 72 per cent of underground rail trips, are journeys to/ from work. I have discussed this feature in greater detail elsewhere (White, 1976), but in brief we can see that the ratio of work to other types of trip rises with city size. This in turn requires a greater ratio of peak to off-peak capacity, with resulting diseconomies. Whereas in a town of about 100,000 the number of buses in service may vary little during the 'working day' (approximately 0800 to 1800), with a lower service level in the evening, giving a peak to inter-peak ratio of about 1.3 to 1.5 (the 'inter peak' is the period between the journey to work peaks), this ratio may reach about 2.0 or higher in some conurbations. It is aggravated by the greater length of work trip as city size rises, which coupled with congestion, makes it very difficult for a bus to perform a 'double run' within a peak period, and thus only one substantial load in the peak direction is carried by the peak extra vehicles.

The imbalance is particularly noticeable on radial routes from low-density suburbs, which generate little off-peak travel for public transport, but may produce high flows (usually by rail) to the central area. Workplaces tend to be much more concentrated than shops or other off-peak trip attractors, for which short-distance travel tends to be typical, even in the largest cities.

the peak: inter-peak imbalance on public transport is also affected by car availability. As car use is restricted in city centres, more cars may be left at home, instead of being used for the work trip, thus becoming available for inter-peak trips such as shopping. With higher activity rates, more shopping may be performed by those in work as a secondary trip purpose (near the office at lunchtime, or en route to home in the evening), rather than separate shopping trips being made in the inter-peak period.

Railways tend to take much of the peak flow to the central business area in large cities, being much better adapted for this task than buses — the segregated track avoids congested streets, and higher capacity can be provided. This reduces some of the imbalance in demand on buses, but presents the railways themselves with a very peaked demand, as in the London case.

The market share estimates shown for public transport in Tables 7.1 and 7.2 are dependent upon occasional surveys. A more up-to-date indication of the use of public transport may be obtained from dividing the catchment population into trips made. For British urban areas in 1980-1, average public transport trip rates were approximately as follows:

100,000 to 150,000 population	100 − 150
150,000 to 500,000 population	125 − 250
conurbations	150 − 300

Within this pattern, wide variation exists, with higher-density older cities often displaying rates above the group average, at around 250-300 (Edinburgh, Newcastle, etc).

Higher levels of public transport use in larger cities also result from the more positive political approach adopted as a result of its greater initial importance. This results in investment grants for new infrastructure (especially underground railways), lower fare levels, and more extensive concessionary fares to groups such as pensioners, all of which increase the trip rate.

Self-contained urban rail systems (as distinct from suburban services as part of the main-line, surface network) have been characteristic of the largest western cities (Paris, London, New York, etc) since the turn of the century, but have also been adopted in smaller cities, as outward growth has occurred, and investment levels increased. Thus cities of about one million upward, such as Stockholm, Lyon, Atlanta and Nürnberg, now possess metros.

International Comparisons

Although Britain forms a convenient starting point, it should not necessarily be taken as typical of Western Europe and North America, especially within the public transport sector.

In those latter areas more rapid economic growth, especially in the most recent years, has produced higher car ownership levels of about 0.33 to 0.40 per capita at national level, and somewhat lower in urban areas, compared with a British national average of about 0.30. In addition, the activity rates may be somewhat higher, especially in view of the present depth of recession in Britain, increasing demand for transport in general. Urbanisation has generally occurred later, resulting in a growth in urban population, except within some older centres, and the opportunity to shape new development around transport links, especially urban railways and rapid transit. Densities are often higher, favouring public transport.

Similar comparisons in the form of the market shares shown in Tables 7.1 and 7.2 for Britain are not easily made, in the absence of comparable data, but cruder trip rate per head indicators can be used. Broadly speaking, public transport trips per head tended to be lower elsewhere in Western Europe than in Britain, but this is no longer the case, with 250 to 300 per annum, or even higher, typical of many larger cities of 500,000 upward, whereas similar British cities have declined to rates as low as 160. In France, in particular, trip rates have increased rapidly in cities such as Lyon, from a previously low level. Particularly high rates are found in Scandinavian, Swiss and German cities, at over 300 per annum.

North America, with a very much higher car ownership rate, and very low suburban population densities, naturally exhibits a much lower per capita public transport trip rate, at around 50 in many urban areas, although rising to 150–300 in large, high density cities with good urban rail networks, such as Toronto, and New York.

Britain still compares quite favourably in terms of the trip rate found in small to medium-sized centres, of up to about 150,000 population, as a result of traditionally intensive bus networks, offering high accessibility for short trips. Marked international differences also exist in trends over time. In Britain, public transport has generally declined in almost all areas since the peak of use in the early 1950s, apart from rail commuting in larger centres. A similar general decline was evident elsewhere in many European and North American cities until around 1970, although this was less evident in those with high rail investment such as Stockholm and Toronto. Since 1970, however, the general pattern outside Britain has been one of increase, firmly reversing earlier trends. Data for London and Paris show how the parallel trends of the early 1970s are followed by a 'scissors movement' in which Paris first equals the London level of public transport trips in 1977, then rises above it.

Such trends are explained by a number of factors:

(a)   Higher levels of investment, especially in underground railways and light rapid transit.

(b)   High levels of operating support. Typically, fares are expected to cover only about 30 to 65 per cent of costs, compared with 70 per cent upward in most British cities (Allen, 1981). In real terms, fares often fell during the 1970s, compared with increases of about 25 to 50 per cent in many British systems.

(c)   In most cases, higher service quality and reliability, associated with simplified ticket systems (reducing delay at stops), fewer industrial disputes, more priority lanes for public transport, etc.

These differences also tend to result in higher productivity (in terms of passenger trips or passenger-km per member of staff). Obviously, there are limits to how far financial support can continue to rise, and some real fare increases are now taking place. Nonetheless, the picture is likely to remain much more positive than in Britain for the foreseeable future, and challenges the assumption that higher car ownership and greater incomes necessarily result in the decline of public transport.

Japan compares similarly with Western Europe, but offering even higher levels of rail use. Private suburban railways play a major role in addition to state and municipal networks.

The Supply of Transport

Finance and Pricing Policy

Transport provision is paid for by two groups:

(a) The users: purchase and operation of cars, cycles, etc. Parking charges. Public transport fares.
(b) Public authorities: road construction and maintenance, street lighting, etc. Support to public transport investment and operation.

In addition, some users have costs paid on their behalf. Employers may provide free parking, and free buses for staff. Cars for 'business' use may be provided not only for those needing them in the course of work, but as a 'perk' in lieu of additional salary. Certain groups such as the elderly and disabled often receive low-priced or free public transport use, with the operator compensated for loss of fare revenue. School children travelling above certain distances (in Britain, two or three miles, depending on age), may be given free travel, either in special-purpose school buses or passes for use on the public network.

Payments made by users may also include substantial taxes, causing the 'cost' they perceive to be above the resource cost. Examples include fuel tax, annual duty on cars, and general taxes such as VAT. The extent to which taxes paid by car users should be seen as a payment for the costs of road use, or simply as a form of general taxation is a matter of opinion, but even as payment for costs they are not a very good proxy. Annual duty on a car does not vary with mileage, but reassigning this to an equivalent duty on petrol would correspond much more closely with use made of the road system.

User perception of cost is also affected by taxation policies. In some countries, the cost of home to work commuting may be offset against income before tax (Westminster, 1980), in some cases based on the equivalent public transport fare. In Britain this is not the case, but those provided with cars for 'business' use have benefited substantially, from the largely non-taxable nature of this benefit (now to be taxed more strictly). Japan illustrates the exact reverse of the British pattern, with very little 'business' car provision, but employers increasing wages by sums needed to cover the cost of public transport commuting, only partially taxed.

Within the local authority public sector, expenditure has traditionally concentrated on road investment and maintenance, but the share taken by public transport has risen rapidly, initially through investment grants (mainly for new railways), and more recently in operating support. Taking approved expenditure for 1982–3, in England, about 20 per cent is for public transport operating support, 48 per cent for road maintenance, and 32 per cent for capital investment in roads and public transport (mostly the former). Per capita public transport expenditure is much higher in the conurbations (£15), compared with the non-metropolitan counties (about £3), contrary to what one might expect. This is because of much higher service levels, and lower fares, in the conurbations, plus capital investment which is negligible in the shires.

So far as the car user is concerned, the cost most easily perceived is that of fuel, and parking, but this is only about 30 per cent or less of total annual costs, when vehicle depreciation, tax, etc. are considered. The cost perceived by the public transport user is related much more closely to the number of trips made, and distance covered, especially where traditional cash-paid ticket systems are in use, with separate payment for each trip, and a fare scale linked to distance.

However, if the aim is to reflect costs of provision, distance alone is not the best guide. About 70 per cent of costs are labour, paid on a time basis, and 15–20 per cent are determined by the peak capacity provided (number of vehicles, scale of depots, etc. – for rail operation the former may be smaller, the latter larger). Only about 10–15 per cent of costs are directly related to vehicle miles run, such as fuel. The importance of the peak suggests that this should be one of the main elements in cost allocation, especially in larger cities where the ratio of peak to inter-peak demand is greater (q.v.). In general, higher fares should be charged at the peak periods. Rather than having scales finely graduated by distance, flat fares (in small towns) or simple zoning systems are more appropriate, especially as these lend themselves to forms of prepayment, simplifying operation considerably (either in the form of 'multiride' tickets cancelled for each trip, or 'travelcards', permitting unlimited use within a defined period, typically one month) (White, 1981b).

System Management

The concept of 'managing' the transport system as a whole — in contrast to operational management within the public transport operators — has attracted increasing attention. Sometimes known by the American term 'Transportation System Management' (TSM) it comprises the construction of limited new infrastructure, traffic management of road space, and public transport innovations to give the best results largely from existing resources through a 'package' of measures, instead of each mode being considered in isolation (Black, 1981).

Considering only vehicular road traffic for the moment, it can be seen that application of traffic management measures has had considerable effect in enabling a road network of a given size to handle substantially greater flows — by about 10 to 20 per cent — than considered practicable about twenty years ago. Means adopted include:

(a) Channelisation of traffic, especially at intersections, for example segregation of left- and right-turning flows.

(b) Introduction of more traffic-light-controlled junctions, and mini-roundabouts, increasing the flow that can be handled by a single intersection.

(c) One-way systems, increasing flow along sections of carriageway of given width, and greatly reducing conflicting movements. at junctions.

(d) Linking of traffic signals with computer control, now known as Urban Traffic Control (UTC).

(e) Limitation of parking, especially on-street parking likely to obstruct flow. In UTC systems, off-street parking sites can now be linked to remote displays, enabling advance warning to be given of full occupancy.

These techniques have been applied widely, especially within Britain, enabling substantially greater traffic flows to be handled within a similar volume of road space, and without average speeds falling (Dept. of Transport, 1978). They are, however, subject to several limitations. Increased traffic flows in formerly quiet side roads, often as part of one-way systems, naturally result in protests from those directly affected. Conditions for pedestrians and cyclists may be worsened, as well as those of bus users where buses are diverted from their principal passenger stops. In addition, regularly relying on traffic flows of the maximum possible size on existing networks may give very little reserve capacity for poor weather or accidents.

Simply raising road network capacity in terms of the number of vehicles may not be the best long-run response to current problems. If accommodating more private cars — with an average load in urban use of about 1.3 — worsens conditions for cyclists and bus users, then these more space-efficient modes may lose users to the private car (in addition to losses which may occur anyway as car ownership rises). The overall result may be that, although vehicular flow is increased, the total flow (as persons) falls. This can be seen in the case of central London. The annual cordon survey (London Transport, 1980), shows that in 1975, 148,000 occupants of buses and 166,000 occupants of cars entered the central area between 0700 and 1000, a total of 314,000 (plus a somewhat larger volume by rail). The number of buses was 3,200, and of cars, 118,000 — a total of 121,000 vehicles. In 1980 the corresponding survey showed 2,600 buses and 137,600 cars entering — a total of 140,200 vehicles. However, occupants had fallen to 287,000, of which 184,000 were in cars and 103,000 in buses. The total number of vehicles rose by 15.6 per cent, but the number of people entering fell by 8.6 per cent. Apart from some growth in cycling, road space overall was being used less efficiently, despite traffic management measures. Energy efficiency was also reduced.

It has already become the practice in 'conventional' traffic management techniques, to provide some segregation of buses, typically in the form of bus-only lanes approaching junctions which, by enabling buses to pass through on the first available green phase reduce both average delay and variation in running times, aiding service reliability. Other simple examples include contra-flow lanes, which enable buses to continue using in both directions a street otherwise converted to one-way flow, and exemptions for buses from some restrictions, such as banned right turns (Vuchic, 1981). These minor priorities do not necessarily cause significant delay to other traffic, but may be seen as an extension of the principle of 'channelisation'. Even if no car occupants are diverted, benefits are given to bus users and operators in time savings.

This concept may be extended to that of giving priority to all 'high occupancy vehicles' (HOVs), which in addition to buses may include street trams (where retained), minibus pools for commuting, and shared cars or taxis carrying about four or more occupants. A similar number of people may be carried in substantially fewer vehicles, permitting a faster flow of traffic, and savings in time and energy. Such methods have attracted the greatest attention in the USA (although it must be said that the very high level of car use there gives the greatest need for them, other countries possessing more extensive public transport systems). Measures taken include creation of very substantial bus priority lanes on urban motorways — up to eleven miles (17.7 km) long, such as the Shirley Highway in Washington, DC, on which a.m. peak public transport patronage quadrupled to reach 16,100 trips in 1974, rising to 21,600 in 1979 (UMTA, 1981). Similar priority may be given for 'van pooling' (i.e. minibuses shared for commuting, driven by employees), and car sharing. Park and ride, in which cars are left at a rail station, or site served by special express bus service, is another element in the package.

In Britain, the 'TSM' process has not been identified as clearly as in America, but the mixture of measures taken often attains similar effects. Given the higher initial level of bus use, the outcome has not been to return existing car users back to higher-occupancy modes, but to assist in retaining fairly high levels of public transport use. Some limitation on off-street car parking, usually by charging a high price for all-day parking (i.e. that used by peak-period commuters) is also characteristic of such approaches, although limited by the proportion of parking (often under half) which is under public authority control. Examples include Reading, Oxford and Southampton, with extensive bus priorities, both on radial corridors and within the central area, some limits on car parking, and in Oxford's case, a successful bus park and ride service used by about 1,500 cars per day.

Encouragement of car sharing has the merit that extra investment in, and operating costs of, additional peak-period public transport to accommodate diverted car users may not be required, the existing car 'fleet' being used more intensively. Some American schemes have included considerable promotion of this concept, but the proportion of users regularly participating is small, around 2 per cent of all existing car commuters (Bonsall, 1981). Simulations of such policies in Britain suggested a very similar result, and the very small (virtually unmeasurable) impact of encouragement given to car sharing under the 1980 Transport Act in Britain bears this out. Furthermore simulation suggests that many of the new car 'sharers' would in fact be former public transport users taking paid lifts, rather than car drivers giving up their own vehicles to ride in others'.

The measures described above go some way to matching road space with demand, but are limited in that a 'carrot' of possibly lower costs and higher speeds is offered, but without any 'stick' aimed directly at the low occupancy peak car trip. Just as use of public transport, and in some cases, car parking, is regulated by price, the same approach may be taken for peak car use. The concept has been discussed for twenty years (and is once again being examined for London), but only one city has implemented such a scheme, Singapore (Wang and Tan, 1981). Since 1975 morning peak car and taxi trips have had to pay a toll of 4 Singapore dollars (about £1) per day, by displaying a windscreen sticker, on entering the central area between 0730 and 1015. Car and taxi vehicular trips have fallen by about 75 per cent, with roughly one-third switching to conventional buses, one-third to car pools (no charge is levied on cars with at least four occupants), and one-third to early travel before 0730. Although some problems remain (notably the lack of a corresponding method of controlling the p.m. peak), the Singapore case certainly shows that such methods can work, given the determination to use them (Holland and Watson, 1977).

So far as the private car is concerned, no fundamental design changes are likely. The performance and fuel efficiency of new models is improving, enabling greater traffic flows to be handled within similar volumes of road space and energy demand. In the British case, it is anticipated that total oil-based energy use for land transport in the year 2000 will be the same as in 1979 (Advisory Council on Energy Conservation, 1981). This estimate assumes a lower rate of growth in vehicular traffic than previously assumed, and a gain in miles per gallon of about 25 per cent for cars, 30 per cent for light vans and 15 per cent for buses. A switch from petrol to diesel for light vans, and more electric traction in the public transport sector is also envisaged. The extent to which average mpg for cars does improve is limited in that, although the mpg for any given size of car is increasing, a switch to larger cars has offset this, such that during the 1970s in Britain, average fuel consumption did not vary significantly from 30 mpg (Rice, 1982). Reduction in, and possible elimination of, lead content in petrol may also offset potential gains in fuel efficiency, although highly desirable on other grounds.

The overall speeds attained by private cars are likely to be determined by road space available, and traffic management. Further growth in car numbers in already congested cities may lead to falls in average speed, and within residential areas deliberate reduction of speed (through measures such as humps) is often desirable. Large-scale road building programmes are unlikely, although some increase in capacity, especially at intersections, is probable.

Within the public sector, greater scope exists for improvement. Boarding and alighting can and should be made easier by reduced step and floor heights, and matching platform heights to vehicle floor heights, as in the busway concept developed by Volvo in Halmstad, southern Sweden. Energy efficiency may be improved substantially by storage of kinetic energy normally wasted in the braking process, notably through regenerative braking on electric railways (White, 1982). Cuts in weight of rail rolling stock give further gains.

Where heavy flows are handled by buses on congested road networks, scope exists for substantial improvements in speed, service quality and efficiency by creating new reserved-track public transport routes. Traditionally, this has taken the form of new railways, usually underground, at any rate in city centres and inner suburbs. Scope for further development of this mode is limited by the very high costs of tunnelling, and lower densities of traffic found in 'western' cities as urban activities become more dispersed. Increasing interest is now shown in light rapid transit, making use of electric rail technology, but with much simpler, lighter structures and smaller vehicles. Short tunnel sections may be found, but a return to street running, in pedestrian/public transport reserved zones, may now be seen. As well as the upgrading of former tramways to LRT standards (and conversion of a former suburban rail network, the Tyne & Wear Metro, to 'heavy' LRT), entirely new LRT systems have been built in San Diego and Buffalo in the USA, Edmonton and Calgary in Canada, and Utrecht in Holland.

Automated systems can be regarded as a further step in LRT technology. Elimination of the driver not only cuts direct operating costs, but enables a much higher frequency of service to be operated, without additional energy or labour costs. Smaller cars may be accommodated in tracks also with a reduced cross-section, requiring less space for new alignments, and automated control can give energy savings through matching acceleration and deceleration phases in different vehicles to make the best use of regeneration. In the early 1970s concepts such as the British 'cabtrack', or American PRT (Personalised Rapid Transit) envisaged very small, car-size vehicles, with individual routeing over an automatic guideway network. However, the complexity and limited capacity of such systems has led to a more practical approach in which automated systems more closely resemble traditional LRT, but with the advantages described above. The first systems came into use in somewhat artificial environments catering for a specialised market, within airports (as will be the case with the first two British systems, at Birmingham and Gatwick), and a university campus (Morgantown, W. Virginia). The first automated systems to cater for a full range of demands, including the residential-workplace link, opened in 1981 in Kobe and Osaka. They link newly-built zones with interchanges on existing rail networks, and operate with short trains of 3–4 cars, each holding about 70 passengers. A more ambitious route began operating in Lille in 1982, the VAL, to be extended further in 1983.

A cheaper alternative, with wider application, is the guided busway. By placing buses on a reserved track, they can be given many of the advantages that urban railways and new automated systems attain (freedom from traffic congestion, etc), but at much lower capital cost. An incremental approach may be adopted, in which sections of busway are built as funds permit, and urban development (or redevelopment) occurs. The simplest form is a two-lane single carriageway built for buses only, as in the largest system to date, in Runcorn New Town. If lateral guidance is provided, then a narrower track may be followed, and automated operation perhaps introduced over the guideway section. This is the basis of the 'O bahn' developed by Mercedes, of which the first section was built in Essen in 1980 (Yearsley, 1981). An entire suburb-city centre radial corridor of 11.8 km is to be served by this means in Adelaide, where the O-bahn was selected in preference to a previous LRT project (being about half the capital cost), construction commencing in February 1982. Articulated single-deckers are to be used.

Little change is likely in the methods of traction used — battery power possessing very little scope save for delivery vans — but further shift from diesel to electric traction should occur as a result of further rail/LRT investment, and a return in some cities to trolleybus working, together with growth of automated systems.

Overall, technological change will improve the present efficiency of the private car, and enable public transport to play its major role on radial corridors more effectively, although doing little to solve problems in areas of low-density demand. Planned concentration of urban development continues to be desirable if sufficiently dense markets to justify new investment are to be found.

Changes in Demand for Travel

Average journey length has risen, as densities have become lower, and urban areas expanded outward, probably at about 1 per cent per year. Shorter trips have reduced in frequency as shops, schools and work-places are no longer within walking distance, and on the public transport network, some shorter trips have been discouraged by reduced frequencies and high minimum fares. Average trip length also rises as a result of people taking advantage of higher speed of modes now available (car, improved rail services) to live at greater distances from activities. This is seen most traditionally in the outward movement of homes from city centres, with continuing radial commuting into those centres over greater distances, but can also be seen in many more dispersed trip patterns — frequent travel to a wider range of friends and relatives, and the ability to switch jobs or places of education without moving home. The latter may be termed 'inertia commuting', the commuting distance being increased as a result of an unwillingness or inability to change house when changing other activities. This may be seen as an extension of personal freedom, with improved transport facilities enabling changes to be made without moving home. However, difficulty in the latter may also be a reason (especially in Britain, with its emphasis on owner-occupation, and council housing).

Population growth in Western Europe has become much slower, with a virtually static total at present, and low birth rate. The rate of growth in North America has also reduced. Older cities have declining populations, falling by as much as 1 per cent per annum, reducing total travel demand within them. In contrast, many cities within developing countries experience rapid growth, as a result of both national population increase, and urbanisation.

Within fairly stable total populations, changes in travel thus come about largely as a result of changes in the trip rate per head. The motorised trip rate has tended to rise, although this may exaggerate growth in total trips, due to former walking and cycling trips, not fully assessed in previous surveys, switching to motorised modes. Trip rates have also risen as a result of rising incomes, an increase in the proportion of the population in work (until recently), and/or education.

There are some exceptions to this. The 5½ day working week has been replaced by a 5 day week, and cinema audiences are about one-tenth of their postwar peak, being replaced largely by television. The increased proportion of elderly within the population also reduces the average trip rate.

The highly peaked pattern of demand during the 'working day' shows welcoming signs of flattening out, such a trend being evident on public transport in Britain for several years. Fewer peak trips are made, as a result of increased unemployment, and inter-peak travel is increasing, notably in trips made by pensioners.

As indicated at the start of this chapter, catering for motorised peak travel demand is receiving less emphasis than before. Permanent changes in employment patterns have occurred, quite apart from short-run effects of the present recession. Large-scale industrial employment is likely to be less important, with service industries playing a greater role. Peaking of demand during the day should continue to diminish, especially if flexitime can be given more encouragement. Increased numbers in further education and training to some extent replace work trips, but here also a better spread of travelling times is likely. Falling school rolls also reduce peak demand.

The length of work trips, with their associated time and cost, suggests that we should try to locate homes closer to workplaces, or vice versa. This has been the basis of much planning effort in the past, notably in British new towns, and to some extent has been successful in creating self-contained areas. Within existing urban areas, many cases now exist where in theory largely residential zones might accommodate more jobs, the greater 'localisation' of employment reducing the length of work trips, and peak demand for infrastructure. One of the most effective means to this end may not be through transport policy, but in the housing field enabling workers to shift home more readily to match changes in workplace. The ability to transfer easily within the local authority housing sector, and less cumbersome means of buying and selling homes could help considerably.

However, the scope for 'localising' workplaces should not be over-stated. Home location is often a product of several factors – the places of work and education of several members, not just the head of house-hold, and access to schools, shops, open space, etc. Individuals may be willing to tolerate seemingly illogical work trips for this reason. Furthermore, employment is becoming increasingly specialised. The extensive semi-skilled/unskilled market which existed in the past is contracting. The probability of someone being able to get a job in the same zone with the skills they possess already, is becoming lower. This applies not only to manual work, but increasingly to office work also. Studies of proposals to shift employment within Stockholm, from the central area, undertaken as part of an International Study (TRRL, 1980) suggest that even large-scale shifting of employment to residential areas would do little to shorten average work trip length. A more dispersed pattern, rather than fewer, or shorter, trips would emerge.

It has also been suggested that certain other activities could be localised, to reduce demand for travel. The clearest example of this lies in entertainment, as television has expanded its role. Otherwise, the scope is questionable. The number of shops will probably continue to fall. Patterns of friends and relatives will continue to become more widespread, with correspondingly longer trips. However, a switch back to smaller schools might be practicable, enabling more pupils to travel on foot or cycle, reducing peak motorised travel demand. There is, however, very little evidence as yet of telecommunications replacing face-to-face contact.

The overall pattern is thus very difficult to judge. The most likely is a roughly similar volume of travel to that found now — with the public transport share highly dependent upon price and quality of service provided — and a somewhat better spread through the day, aiding utilisation of vehicles and infrastructure.

References

Advisory Council on Energy Conservation (1981) *Review of the UK Transport Energy Outlook and Policy Recommendations*, Department of Energy, London, HMSO

Allen, J.E. (1981) 'Fares — Who Pays?', *Transport* (Journal of the Chartered Institute of Transport, London) *2 (6)*, 53–4

Bonsall, P.W. (1981) 'Car Sharing in the United Kingdom', *Journal of Transport Economics and Policy*, *15 (1)*, 35–44

Black, J. (1981) *Urban Transport Planning*, Croom Helm, London, Chs. 8 and 9

Goodwin, P.B. (1980) 'The Incorporation of Walking in Transport Methodology', Conference on Walking, Policy Studies Institute, London

Holland, E.P. and Watson, P.L. 1977) 'Measuring the Impacts of Singapore's Area Scheme', World Conference on Transport Research, Rotterdam

Jones, P.M., Dix, M.C., Clarke, M.I. and Heggie, I.G. (1980) *Understanding Travel Behaviour* (Transport Studies Unit, Oxford)

London Transport (1981) *Annual Report and Accounts 1980*

Maltby, D., Lawler, K.A. and Monteath, I.G. (1978) 'A Monitoring Study of Rail Commuting on Merseyside 1974 to 1978', *Traffic Engineering & Control*, *19 (6)*, 278–82

Nash, C.A. and Johnson, I. (1981) 'Preliminary Results from a Survey of Present and Past Rail Commuters in the Hertfordshire Area', Leeds University Institute for Transport Studies, working Paper 151

National Travel Survey (1979) *National Travel Survey 1975/6 Report*, HMSO, London

Pacione, M. (1981) *Problems and Planning in Third World Cities*, Croom Helm, London

Rice, P. (1982) 'Trends in GB Fuel Efficiency (1970–80)' *Traffic Engineering and Control*, *23 (4)*, 224–8

Transport and Road Research Lab (1980) *The Demand for Public Transport: An International Collaborative Study*, TRRL, Crowthorne

Transport Statistics Great Britain (1970–80) HMSO, London 1981

Urban Mass Transit Administration (1981) *Traveler Response to Transportation System Changes*, 2nd edn, Washington, DC

Vuchic, V.R. (1981) *Urban Public Transportation Systems and Technology*, Prentice-Hall, New Jersey, section 4.3

Wang, L.H. and Tan, T.M. (1981) 'Singapore', in M. Pacione (ed.) *Problems and Planning in Third World Cities*, Croom Helm, London, pp. 218–49

Westminster Chamber of Commerce (1980) *Travelling to Work*, London, pp. 31–3

White, P.R. (1976) *Planning for Public Transport*, Hutchinson, London

White, P.R. (1981a) 'An Evaluation of the Long-term Effects of the West Midlands Travelcard', West Midlands PTE 1981 (unpublished)

White, P.R. (1981b) 'Recent Developments in the Pricing of Local Public Transport Services', *Transport Reviews*, 1 (2), 127–50

White, P.R. (1982) 'Energy Conservation in Urban Public Transport', *Transportation Planning and Technology* (to be published)

Yearsley, I. (1981) 'Where are Guided Buses Taking Us?', *Modern Tramway* (London), *44 (523)*, 218–23

8 HEALTH

J.A. Giggs

Many people believe that the health of city dwellers has always been worse than that of rural populations. It has long been argued that urbanisation and urban living conditions have had adverse effects upon human health and behaviour (Harrison and Gibson, 1976; Moos, 1976; Weinstein, 1980). Unfortunately, few reliable historical data are available which indicate variations in illness and mortality between urban and rural populations. Since the mid-nineteenth century however, a modest amount of epidemiological research in Western countries has produced evidence which suggests that urban environments contribute both to the origin and severity of many physical diseases. For behavioural (that is mental) disorders though, the evidence is conflicting, owing to substantial temporal and regional variations in diagnostic methods, treatment practices and availability of facilities (Kaplain, 1971; Srole, 1972; Giggs, 1980). More importantly, the available epidemiological evidence reveals that there are often profound spatial variations in morbidity and mortality levels among populations living within large urban areas.

During the past thirty years geographers have become increasingly concerned with several aspects of urban health variations. Their researches form a relatively small portion of the broader field of medical geography and their perspectives have been greatly influenced by increasingly varied technical, methodological and conceptual developments in that field. The major developments within medical geography have been comprehensively documented in recent reviews by Learmouth (1975), Pyle (1979) and Phillips (1981).

The present chapter provides a review of the main conceptual approaches and empirical research specifically at the intra-urban scale. Four major areas of research are discussed here: disease mapping, ecological associative analysis, disease diffusion studies and the geography of health care. They are not given equal consideration here. However, this should not be interpreted as an indication of their relative importance. Thus although disease mapping and associative analyses have long been the central concerns of medical geography the intra-urban literature on these topics is not reviewed at length in the present chapter because this has already been done elsewhere (for example, Herbert, 1976; Giggs, 1979, 1980). In contrast, the newer conceptual approaches have been relatively poorly served. This chapter therefore deliberately concentrates upon developments in disease diffusion studies and the geography of health care. The latter subject is particularly important because it incorporates several fairly distinct research foci — notably the structure and spatial patterning of medical services, health services utilisation (incorporating consumer behavioural studies) and health service planning studies.

Mapping Ill Health and Mortality

Medical cartography has long been an important part of medical geography and has contributed much to our understanding of the spatial aspects of human health problems. Thus Gilbert (1958, 1972) has shown how several Victorian pioneers used dot maps to identify the link between cholera incidence and polluted drinking water supplies in a number of English cities. Ford (1976) has reviewed similar studies of cities in the USA during the nineteenth century. This simple carto-graphic technique is still being used effectively, particularly in the analysis of contagious disease diffusion. Intra-urban examples include a measles epidemic in Akron (Pyle, 1973) and a cholera epidemic in Braganca Paulista (Morrill and Angulo, 1979). The tedious work of locating and plotting numerous individual cases is now being superseded by the development of automated patient information files and geo-graphic (that is small area census) base files. Pyle (1979) has described how these files have been used to produce automated maps of the distribution of gonorrhoea tests for black and white persons for city wide and street block scales in Columbia, South Carolina.

Several other traditional cartographic methods have been used in mapping health problems and mortality in cities (Pyle, 1979). Crude and age standardised rates have most commonly been employed in these analyses. These have been calculated for the relevant populations living in urban sub-areas which range in size from street blocks (for example, Dever, 1972) to Metropolitan Boroughs (for example, Sainsbury, 1955). Several important methodological and conceptual problems are raised in this kind of mapping, notably the effects of contrasting sub-area/population sizes (Girt, 1972), population structure (Wilson, 1978; Keig and McAlpine, 1980), population turnover (Burnley, 1977) and the familiar dangers of ecological interpretation.

Figure 8.1: Distribution of Poverty Areas and the Early Generations of a Measles Epidemic in Akron

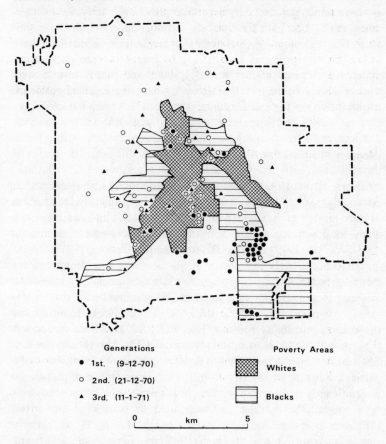

Source: Pyle, 1973.

Nevertheless, the results of most studies have demonstrated that there are marked variations in the spatial distribution of most forms of ill health and mortality within large cities. These analyses have often prompted more detailed 'causal' research within high risk areas or attracted the attention of public health officials.

Ideally disease mapping should be repeated at regular intervals. In this way the medical geographer can determine whether a particular disease distribution is relatively ephemeral or persistent. Unfortunately, temporal studies are rare, chiefly because of the considerable effort involved in collecting and analysing the relevant morbidity and mortality statistics. These problems are compounded by the fact that accurate census statistics only become available every ten years. Furthermore, in the UK, both the numbers and boundaries of the urban sub-areas (for example electoral wards and enumeration districts) have varied for every census since 1961. In future the use of annually updated automated patient record systems and census data for grid squares should overcome these currently daunting practical problems. In the meantime the existing findings confirm that temporal studies are of great value. At one extreme, Pyle (1976) has mapped the changing distribution of fourteen generations of a measles epidemic which lasted barely eight months in Akron (Figure 8.1). At the other, the results of three successive studies of suicide in London (Sainsbury, 1955; Whitlock, 1973 and Howe, 1979) show that the distribution of high rate areas has remained remarkably stable over a period of nearly 60 years. Information of this kind is essential to public health planners and has prompted many local authorities to establish automated area-based files in both the USA (Pyle, 1979) and the UK (for example, Webber, 1975).

For comparatively rare forms of disease there is a strong case for mapping probabilities rather than absolute frequencies, or rates. The evidence to support this view has come from a number of studies at a variety of spatial scales (McGlashan, 1976; Pyle, 1979). In most cases the Poisson probability formula (Norcliffe, 1977) has been used to test the variability of human morbidity or mortality for selected diseases. The test enables the researcher to determine whether the incidence of a particular disease in any of the urban sub-areas (for example, wards) is significantly higher or lower than in the population of the study area as a whole. The method has been used in studies of intra-urban variations in cancer mortality in Pittsburgh (Patno, 1954), primary acute pancreatitis morbidity in Nottingham (Giggs *et al.*, 1980) and schizophrenia morbidity in Nottingham (Giggs, 1982).

Environmental Health Hazards and Associative Analyses

A second important field within medical geography has been the development of what have been described as 'associative analyses' (McGlashan, 1972; Pyle, 1979). Here the focus of attention shifts from the mapping of disease patterns to a concern with the relationships between the observed patterns and relevant components of man's environment. This research has been deliberately described as 'associative analysis' because it tends to offer statistical evidence, rather than absolute proof, of 'causal' relationships between the distributions of specific environmental phenomena and particular diseases.

The explanatory limitations can partly be ascribed to research design, for most of the work in this field has been conducted at the ecological (that is aggregate) level. Moreover, virtually all the research published to date has been static, or 'period picture' in character. However, clinical and behavioural studies indicate that, for most physical and mental disorders, pathogenesis can rarely be ascribed to a single environmental factor. In most cases morbidity and mortality are functions of several interacting phenomena, often operating over fairly long periods of time. Even when all these inimical factors can be identified, it is rarely the case that appropriate (that is quantified) data exist, or that their relative significance can be assessed (Kasl, 1977; Smith, 1977, 1980; Wood *et al.*, 1974; Wood and Lawrence, 1980). In these circumstances, therefore, correlation of disease-environmental factors at the time when cases are first identified (for example, by general practitioners) may yield inaccurate results.

In exceptional cases the interaction between environmental hazard (that is cause) and disease morbidity and mortality (that is effect) operates quickly and dramatically. One of the best documented examples was the great London 'smog' of 5–9 December 1952 (Ministry of Health, 1954; Wilkins, 1954; Martin, 1964; Royal College of Physicians, 1970). Cold foggy anticyclonic weather conditions produced a prolonged temperature inversion and a massive build-up of atmospheric pollution (chiefly smoke and sulphur dioxide) in London. The resulting smog was probably responsible for much illness and hastened the deaths of 3,500 to 4,000 vulnerable people over and above the number normally expected for the period. Most of the increases in mortality were linked with lung diseases (bronchitis, pneumonia and respiratory tuberculosis) and heart diseases (coronary heart disease and myocardial degeneration). Mortality rates stayed abnormally high for several weeks after the event as the adverse effects of the event took their toll.

It is more commonly the case that the inimical effects of pollution and other environmental hazards on health are subject to considerable time lag. Thus in Greater Manchester Wood and colleagues (1974) found that the 1970 mortality rates for bronchitis, emphysema and all causes correlated strongly with the high average smoke concentration for the winter of 1963–4. In a more recent article (Wood and Lawrence, 1980) the authors describe the background to air pollution control in

Greater Manchester and the resulting falling pollution concentrations. Analysis of the statistical relationships between mortality rates for 1973-5 and three cumulative air pollution data sets covering 1963-75 yielded no significant correlations between mortality rates and smoke concentrations and only weak correlations with sulphur dioxide levels.

These results indicate that air pollution control appears to have reduced the incidence of deaths from diseases attributable to the inimical effects of atmospheric pollutants. However, both of these investigations also contain results which demonstrate the complexity of disease-environmental relationships and the difficulties involved in unravelling the relevant causal and temporal elements. Thus in the first investigation (Wood *et al.*, 1974) the statistical relationships between mortality and socio-economic variables and population density were as strong as those with air pollution variables. In the second study (Wood and Lawrence, 1980) the correlations with social and housing variables were much stronger than those with eight pollution variables. Furthermore, the authors suggest that their social variables are probably only surrogates for more significant behavioural factors (for example, smoking, diet and exercise). Similar conclusions have been drawn in a canonical analysis of mortality data and ecological structure in Houston (Briggs and Leonard, 1977). It would seem therefore, that future improvement in the health status of urban populations will depend primarily upon alterations in the behaviour patterns and lifestyles of individuals and social groups.

The findings of these and many other 'associative' studies of disease environment relationships within urban areas demonstrate that the traditional ecological approach presents many interpretative problems (Giggs, 1979, 1980; Phillips, 1981). In a recent review of the subject Pyle (1979) has indicated that a sound understanding of the merits and limitations of these analyses depends heavily upon better research design and the careful use of a *range* of statistical methods in a properly organised modelling sequence.

In the longer term the value of associative analyses will undoubtedly rest upon the development of new strategies (that is research designs and techniques) and a growing understanding of the relevant clinical and behavioural variables. This research will also require the collaboration of workers in a range of relevant disciplines since it is clearly multivariate in character and beyond the conceptual and technical grasp of individuals trained in particular, specialised, fields. A good example of the potential afforded by this kind of collaborative research is provided by three recent studies of suicide in Brighton. In the first of these the authors

(Jacobson *et al.*, 1976) explored the question of *case identification* by examining the clinical and social variables which differentiated suicide, open and accidental death verdicts given by coroners in Brighton. In the second paper (Bagley *et al.*, 1976) principal components analysis was used in a taxonomic analysis of 123 variables relating to the social, family and clinical circumstances of all identified suicide cases in Brighton. Many of the variables were obtained from detailed interviews of friends and relatives of the deceased persons by qualified interviewers. The analysis suggested that three distinct types of suicide might exist. These were described as 'clinically depressed' (Component I), 'sociopathic' (Component II) and 'old and handicapped' (Component III) suicides. In the third investigation (Bagley and Jacobson, 1976) the authors tested the hypothesis that these three apparently different kinds of suicide varied significantly between three ecologically distinct areas within Brighton. A chi square analysis showed that the 'sociopathic' suicide type occurred chiefly in central Brighton and the 'old and handicapped' suicide type in middle class areas.

Diffusion Studies

It has long been recognised that many human diseases spread out over space and time from specific source areas (for example, Hirsch, 1883-6). However, the systematic analysis of the relevant patterns and mechanisms became a clearly discernible and distinct research field within medical geography only in the late 1960s (Pyle, 1977). Its emergence at that time can be linked with the general growth of diffusion studies within the literature of human geography. Because of the relative novelty of the subject the medical geographical literature on diffusion is still both small and fragmentary. Furthermore it is characterised by a considerable range of approaches and explanatory strategies, many of which require further development (Pyle, 1979, pp. 123-4). Despite all these limitations there is clear evidence that the range of techniques used in diffusion studies frequently offers superior insights into many medical phenomena than those employed in most of the more familiar static (that is 'period picture') analyses.

One of the most critical issues in spatial diffusion studies remains the question of the particular geographical scale to be adopted. In many instances it is not possible to isolate the effects of any one spatial scale. Thus Haggett (1976) showed that the spread of measles in Cornwall between October 1966 and December 1970 could only be 'explained' by processes operating simultaneously at a variety of spatial scales. Recognising these problems, most medical geographers have tended to adopt an empirical approach and have examined the spatial and temporal diffusion of specific medical phenomena over a limited range of spatial scales.

Most of the empirical research to date has been concerned with disease diffusion at the macro-regional (that is international, national and regional levels). In comparison, the amount of geographical work on the diffusion of medical phenomena at the inter- and intra-urban scales has been trifling. This is a curious and regrettable situation, given the fact that the bulk of the populations in developed countries live in urban settlements. Nevertheless, the few existing studies confirm that geographers can make important contributions to our understanding of medical phenomena. A particularly telling point made by Pyle (1979) is the fact that the spatial and temporal analysis of disease epidemics can provide information concerning the locations of outbreaks. This could prompt the early implementation of disease control and prevention measures, with a potentially substantial consequent saving of suffering, lives and resources.

There are very few published geographical studies dealing with the spatial diffusion of medical phenomena at the intra-urban scale. Here the concepts and methods of diffusion research have been applied chiefly to the analysis of infectious diseases. These tend to occur in time series cycles or nearly normal distributions. Consequently traditional ecological analyses tend to obscure or miss potentially significant aspects of the interactions between disease and environment. This point can be exemplified with reference to a meticulous study of measles in Akron, Ohio (Pyle, 1973). The analysis began with the calculation of adjusted age-specific rates of measles for children aged under ten years. These were calculated and mapped for the city's 69 census tracts for the years 1965, 1966, 1969 and 1970. Inspection of the four maps revealed a persistent concentration of the highest rates in the central and south-eastern parts of the city. A comparison of the four 'measles rate' maps and one showing the actual distribution of children under ten years old produced a surprising finding. The relationship was completely inverse, for the tracts with the greatest numbers of children were located on the city's fringe. Pyle then attempted to explain the observed patterns of variation in the incidence of measles in 1970. However, regression analysis of the measles variable and eleven relevant socio-economic and population variables produced only weak to moderate correlation coefficients.

Disaggregation of the measles data provided a solution to the problem. Pyle was able to identify and map fourteen successive generations in the epidemic between late November 1970 and late May 1971. The initial outbreak was reported in a poverty neighbourhood in south-eastern Akron (Figure 8.1). Thereafter a clear pattern of contagious diffusion emerged, with succeeding generations appearing in the inner city and then in intermediate locations. Few cases appeared in the high status suburbs due to the activity of public health inoculation teams. As the epidemic waned the number of cases diminished and consisted mainly of pre-school children living in the locale of the original outbreak. This suggested that older patterns of measles spread persisted in the inner city, despite the availability of vaccine.

Pyle concluded that the measles epidemic raised important issues for health-care policy and behavioural research. Although live measles vaccine had been licenced in 1963, and was readily available thereafter, it was relatively expensive. Consequently price acted as a barrier to the diffusion of inoculations to the entire population 'at risk' and especially to the children of low-income groups living in the inner city neighbourhoods. Between 1964 and 1971 the city's Department of Public Health did not operate a measles inoculation programme in schools. The behavioural pattern of lower income groups with respect to health care treatment also acted as a barrier to inoculation because many parents were either unaware of the existence of alternative treatment facilities, or unwilling to accept vaccination as a protective measure.

Several studies in other cities confirm the merits of specifically geographical analyses of disease diffusion. In Australia, the spread of infectious hepatitis has been investigated in Greater Wollongong-Shellharbour (Brownlea, 1967, 1972) and in Metropolitan Sydney (Baczkowksi, 1980). An outbreak of smallpox in Braganca Paulista, Brazil, has been analysed in great detail by several workers (Klauber and Angulo, 1976; Angulo *et al.*, 1977, 1979; Morrill and Angulo, 1979, 1981). In Ibadan, Nigeria, Adesina (1981) has modelled the spatial and temporal diffusion of cholera. In an interesting historical study, Florin (1971) was able to identify urban to suburban diffusion patterns in average age of death among residents of early-nineteenth-century urban settlements in New England.

Knox (1964, 1971) has argued that an understanding of the causes (that is *aetiologies*) of epidemics of rare diseases is best sought in the study of the ways in which infections spread from one person to another. He reviewed a number of suitable techniques, including network analysis and time-space clustering. These methods are essentially microgeographic and are best used in intra-urban research. Time-space clustering techniques have been used in studies of the distribution of Crohn's disease (Miller *et al.*, 1975) and primary acute pancreatitis (Giggs *et al.*, 1980) in Nottingham.

The Geography of Health Care

During the past twenty years a growing number of geographers and other social scientists have begun to study the spatial aspects of health services delivery systems and their utilisation patterns. Most of this work has been undertaken in highly urbanised developed countries. Here it has been stimulated by an increasing concern over the growing burden of health services expenditure, the effectiveness of the organis-ation of health care delivery systems and spatial inequalities in the provision of services and their accessibility to consumers. The present review is necessarily selective and illustrates the most important methods and problems.

Particular care should be taken to avoid facile comparisons between countries because the health care delivery systems of nations differ in varying degrees. Thus Roemer (1977) has suggested that five broad national types can be recognised: free enterprise, welfare state, transitional state, underdeveloped and socialist state. Some of these types have been examined in greater detail than others. The free enter-prise type is best represented by the USA and comprehensive reviews of developments in the geography of health care in that country have been provided by Shannon and Dever (1974) and Pyle (1979). The welfare state type has been rather more modestly served by geographers. Examples include the UK (Phillips, 1981), New Zealand (Gross, 1974) and, to some extent, Australia (Gross, 1974; Stimson, 1980).

Three main strands can be recognised in the spatial analysis of urban health care. The first strand comprises studies of the various supply components of the medical care system (for example, hospitals, specialists and general practitioners) and their locational determinants. Behavioural approaches to the study of health services utilisation con-stitute the second strand of research. Here the major focus of interest is the consumer and many geographers have analysed the major elements of patient behaviour (and its determinants) in 'consuming' health care services. The third strand is essentially planning oriented. Researchers in this particular field are chiefly interested in resolving the mis-match between the supply and demand for specific health services.

The Intra-urban System of Medical Services

In most countries the medical care system is hierarchical in form and has three tiers — primary physicians (also called general practitioners or GPs), specialist physicians and hospitals. Sick people generally gain access to the system through a GP, although some obtain primary medical care at the out-patients departments of public hospitals. Access to the second tier of the hierarchy — specialist medical services — is generally gained through referral by a GP. The third tier — hospital — is normally reached by referral from a GP or a specialist physician. In some countries (for example, the USA and Australia) these tiers are divided between the private sector and the government's health services (McGlashan, 1977).

Since the early 1960s many geographers have mapped and analysed the spatial distribution patterns of health care service facilities within cities. In some studies traditional theoretical models such as central place theory and location decision choice models have been used to analyse these spatial patterns (Shannon and Dever, 1974; Pyle, 1979; Stimson, 1980, 1981). However, it is now accepted that these methods offer only limited explanations of the functional and spatial relationships that exist both between and within the three levels of the health care delivery system. Their idealised and aggregative assumptions concerning the spatial behaviour of both suppliers and consumers have also presented substantial explanatory problems (Phillips, 1981; Stimson, 1981).

Despite these limitations there is both theoretical and empirical evidence which suggests that it is possible to identify at least some aspects of the hierarchical nature of intra-urban medical care facilities, especially among hospitals. Thus a hierarchical structure has been identified for hospitals in Chicago (Morrill, 1966; Morrill and Earickson, 1968), Cleveland (Shannon *et al.*, 1973) and Adelaide (Cleland *et al.*, 1977). Here several discrete grades were recognised, based upon hospital size (number of beds) and levels of specialisation. In the case of Cleveland, Shannon *et al.* (1973) identified five grades of hospitals ranging from small community units to the major teaching and research centres.

The spatial distribution of the hospital system in large cities broadly resembles that of retail business centres. Typically the higher and middle order hospitals are concentrated in the inner city, whereas lower order units are found primarily in rapidly growing suburban residential areas and in old settlements which have been recently engulfed by the expanding metropolis. However, there is very little suburbanisation of the higher order hospitals and the building of outlying community hospitals frequently lags behind demand for periods ranging between twenty and fifty years (Morrill, 1966; Earickson, 1970; Cleland *et al.*, 1977). The persistent and sometimes intensifying dominance of the major inner city hospitals has been ascribed to inertia of historical location, a preference for on-site expansion, the attachment of specialist physicians to existing prestigious locations and the increasing role of hospitals in medical practice (Morrill, 1966; Miller, 1977).

The cumulative effect of the location decisions and preferences of hospital service suppliers over decades has been to create an increasing spatial and social inbalance between the distribution of services and consumers. As a result most consumers in the outer urban and suburban zones of large cities usually have to travel long distances to gain access to a hospital.

Their difficulties are further compounded in countries like the USA and Australia, where urban hospitals are not equally accessible to all citizens. Many hospitals do not provide free (that is, public) or subsidised services for poor persons (Shannon and Dever, 1974; Stimson, 1980). Furthermore, in the USA many private hospitals cater exclusively for minority groups such as the elderly, negroes, Jews and Roman Catholics. These variegated financial and cultural constraints have created marked intra-urban differentials and barriers in terms of both social and spatial access to necessary hospital services for many minority groups and communities (Morrill, 1966; Earickson, 1970).

Spatial variations in the distribution of specialist physicians – the intermediate tier in the health care system – have been examined by only a few social scientists. Some have applied central place theory to the subject, claiming that specialists can be ranked in a hierarchical fashion according to their levels of specialisation (for example, Dickinson, 1954; Morrill, 1959; de Visé, 1973; Lankford, 1974). However, empirical evidence has shown that the predicted orderly nested spatial patterning of specialists does not occur (for example, Dorsey, 1969; Elesh and Schollaert, 1972; Freestone, 1975; Guzick and Jahiel, 1976; Gober and Gordon, 1980).

The investigation of private practice physician location in the Phoenix metropolitan area is particularly comprehensive (Gober and Gordon, 1980). Fourteen physician types were selected and classified according to their levels of speciality and hospital/community involvement in 1970. The authors used dot maps and standard distance statistics to measure the extent of both office concentration near the city centre and of clustering around hospitals. Although the findings for all groups generally conformed to the expected patterns, they also exhibited higher levels of both concentration and clustering than the expected ideal (that is optimum, market-serving) distributions. The most pronounced deviation was identified in the case of paediatricians. These specialists provide relatively low order services, in that they have a low population threshold (1 per 10,000 population compared with 1 per 100,000 for a thoracic surgeon), they treat children and spend relatively little time working in hospitals. The authors had therefore hypothesised that paediatricians' offices would have dispersed, non-clustered distribution patterns. However, analysis of the actual distribution revealed a very high degree of clustering around hospitals in both central and suburban Phoenix. From the consumer's point of view this was clearly an inappropriate distribution since the majority of the patients (that is children) resided mainly in new housing estates on the periphery of the city.

Little work has been done on the changing spatial distribution of specialist physicians at the intra-urban scale. The evidence provided for US cities indicates that specialists are tending to agglomerate increasingly near the major inner city hospitals, and in some suburban areas, especially around some large suburban hospitals and in planned shopping centres (Eisenberg and Cantwell, 1976; Guzick and Jahiel, 1976; Miller, 1977). The factors which explain specialist locational behaviour are complex, but proximity to patients is evidently not a primary consideration (Shannon and Dever, 1974; Miller, 1977). Ecological and behavioural analyses in US cities indicate that proximity to hospitals, medical research and teaching facilities is a major attraction for practising specialists. Given the growing complexity and volume of medical technology, knowledge and skills, this is scarcely surprising. It is becoming increasingly necessary and attractive (for financial and prestige reasons) to specialise and to work in or near major hospitals. In 1949, 37 per cent of active practitioners in the USA listed themselves as specialists and in 1973 it was 82 per cent (Miller, 1977). The burgeoning costs and complexity of medical care have created an increasingly central role for hospitals in organising medical practice. This phenomenon also explains the growing importance of the hospital outpatient department for ambulatory care at the expense of the 'traditional' GP.

The degree of concentration, or clustering, around hospitals obviously varies with level of speciality and the degree of hospital orientation (Miller, 1977; Gober and Gordon, 1980). However, even among the many specialists who do not practise in full-time capacities from hospitals, there is a trend towards the establishment of multi-speciality group clinics in middle class suburbs. Here they generally have shared office and laboratory assistants. Other important criteria in the determination of practice location for these particular specialists appear to be high median area income (Dorsey, 1969; Robertson, 1970; Guzick and Jahiel, 1976), population growth (Rosenthal, 1978), availability of adequate hospital facilities, acceptance of location to spouse, openness and receptivity of the local medical community (Diseker and Chappell, 1976) and avoidance of areas dominated by minority populations (Miller, 1977).

General practitioners are the major sources of medical care. In Britain they deal with over 90 per cent of cases handled by the entire health services hierarchy (Office of Health Economics, 1974). They thus constitute the basic stratum of the health service hierarchy, providing the most frequently demanded services and referring patients to higher forms of care. Their key role therefore requires high levels of accessibility, that is an even spatial distribution which conforms broadly with that of the total population in a specific area. Numerous studies by many social scientists, however, have shown that uneven spatial distribution (and therefore marked inequality of access opportunity) is typical at national, regional and intra-urban scales in most countries. This situation has been identified in recent literature reviews for the USA (Eisenberg and Cantwell, 1976; Pyle, 1979), the UK (Phillips, 1981), Australia (Stimson, 1980, 1981) and New Zealand (Barnett, 1978).

The evidence provided by temporal studies indicates that the problem of spatial inequity has intensified in most developed countries since the early 1950s. The supply of GPs has not kept pace with population growth because increasing numbers of trainees are tending to specialise. In addition, organisational changes designed to achieve economies of scale have had profound effects upon their locational patterns. In 1951, 80 per cent of the unrestricted GPs in England worked in single handed or two-partner practices. By 1976 this proportion had fallen to 37 per cent. There has clearly been a marked shift in GP location towards group practices and health centres. This trend means that they are now working from far fewer points (that is surgeries) than existed some thirty years ago. At the intra-urban scale the effects of these trends have been profound. Studies in the USA (Dewey, 1973; Pyle, 1979), the UK (Knox, 1978, 1979; Phillips, 1981) and Australia (Stimson, 1980, 1981) show that many communities have experienced substantial losses in patient accessibility to care. Many urban neighbourhoods now

Figure 8.2: Surgery Location and Relative Levels of Accessibility in Aberdeen (1950 and 1973)

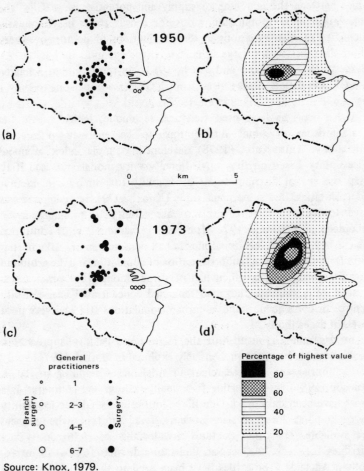

Source: Knox, 1979.

have either no GPs or branch surgeries which open only for limited periods (Figure 8.2 a, c). In most large cities, therefore, there are pronounced spatial variations in both the incidence of primary service provision and in the type of organisation, defined in terms of the size of the unit (that is ranging from the solo surgery to the large health centre) and the quality and availability of service (for example, range of ancillary staff, physical quality of surgeries, hours when GPs are available).

Geographers have used a variety of simple techniques of spatial accessibility analysis to investigate the spatial variations that exist in cities between the matching of supply and potential use of GPs. The calculation of GP:population ratios for census tracts has frequently identified massive variations in the distributions of doctors over very short distances within large urban areas (for example, de Visé, 1973; Barnett, 1978; Bohland and Frech, 1982). In metropolitan Adelaide the GP:population ratios ranged from 1 GP per 448 persons to only 1 per 5,678 in 1977 (Cleland *et al.*, 1977).

Other workers have used statistical techniques which enable the characteristics of point distributions to be analysed with some objectivity. Thus Knox (1978) developed a simple index of nodal accessibility based on the conventional gravity model and used it to map the spatial distribution of accessibility to primary care in four Scottish cities. In a subsequent study (Knox, 1979) the technique was used to identify relative levels of accessibility to primary care in Aberdeen in 1950 and 1973. The results (Figure 8.2 b, d) confirmed that a high level of locational inertia had persisted among GPs in that city. Despite the marked suburbanisation of population and the profound organisational changes which had occurred among GPs over nearly a quarter of a century, they had remained concentrated near the city centre. In consequence, the suburban populations still had very poor levels of accessibility.

In Australia the utility of the relative accessibility approach to health care has been comprehensively explored in Morris's (1976) study of Melbourne and by researchers in Adelaide (Cleland *et al.*, 1977; Stimson, 1981). For Adelaide two small area digitised computer data bases have been developed. One file identifies every GP service facility by geographical location, size, services offered and times when services are available. The other contains a wide range of 1976 census data variables for over 1300 census collector's districts (CCDs) in metropolitan Adelaide. These files have been used to show such important phenomena as temporal variations in the availability of GP services and the degree of spatial congruence between potential GP demand and its potential satisfaction through available GP services for six time periods during the day.

The reasons for the mismatch between surgery location and population distribution within cities have been well documented for the USA and Australia (Shannon and Dever, 1974; Eisenberg and Cantwell, 1976; Pyle, 1978; Australian Medical Association, 1972; Stimson, 1980, 1981). In those countries GPs are primarily influenced by local effective demand, as reflected in high per capita incomes and the ability of the local population to generate (that is to want rather than need) medical care. Other factors include proximity to specialist and hospital facilities, availability of offices in new suburban shopping centres, neighbourhood environmental quality and attractiveness, the social preferences of doctors and their spouses, and ethnicity.

Studies of GPs and their location decisions in the UK suggest that similar factors are at play (Knox, 1978, 1979; Knox and Pacione, 1980; Phillips, 1981). Despite the passing of the National Health Service Act in 1946 and the subsequent reorganisation of the medical services in 1948 equality of access to GPs has scarcely improved. A comparison of the distribution of surgeries in most British cities reveals a high degree of inertia, with considerable 'over-doctoring' in inner city locations (for example, Figure 8.2). In contrast, postwar suburban council and private housing estates have few GPs because of the dearth of suitable accommodation for surgeries (Knox, 1978). City planners, estate developers and builders have been (perhaps unwittingly) responsible for this situation.

Health Services Utilisation

Many models have been developed to analyse the patterns and determinants of patient use of health care services. Most of the early research was concerned specifically with the question of patient accessibility to medical care facilities. Researchers attempted to measure aggregate travel patterns to facilities, employing straight-line distance or travel time as measures of distance travelled between home and supplier. Variants of the gravity and exponential models (Morrill and Earickson, 1969; Morrill et al., 1970; Earickson, 1970; Shannon et al., 1969, 1973), simulation models (Moyes, 1977) and three dimensional analogue models (Bashur et al., 1970; Guptill, 1975; Shannon and Spurlock, 1976) were subsequently used to describe these patterns. In most instances these studies confirmed the existence of a clear distance-decay relationship, for the rate of use of physicians and hospitals generally varied inversely with distance.

For large urban areas these relationships vary in strength and are often complicated by numerous distorting factors. Thus the size, speciality structure and location of such facilities as hospitals and health centres have been found to exert marked influences on the extent, intensity and form of their catchment areas (Pyle, 1979; Phillips, 1981). When large hospitals are highly clustered near city centres their service areas tend to be elongated in the direction of fewest intervening opportunities, that is outwards to the suburban fringe (Schneider, 1967; Shannon *et al*., 1975; Pyle, 1979).

The appearance of such complicating factors in traditional normative analyses have prompted several workers to seek additional frameworks for the explanation of consumer use of health services (Phillips, 1981). In particular, the recent development of a behavioural perspective for the analysis of use patterns and attitudes has contributed substantially to geographical understanding and explanatory power in this field. In this context the evidence presented in the preceding review of research into the major components of the health service system has already shown that the changing behaviour, attitudes and organisation of the various health service suppliers have had profound repercussions upon their patterns of utilisation by consumers.

Although the issue of the attitudes and behaviour of consumers has been comparatively neglected there is sufficient theoretical and empirical evidence to support the view that this should become an important field of research in medical geography during the coming decade. Comprehensive reviews of methodologies and results have recently been provided by Veeder (1975), Thomas (1976), Shannon (1977), and Phillips (1981). Virtually all the geographical empirical research which has been published to date consists of surveys of specific aspects of consumer behaviour *vis-à-vis* particular components of the health care system (for example GPs, specialists or hospitals). In contrast the few existing surveys of consumer health care behaviour across the whole spectrum of the health service system (for example Krupinski and Stoller, 1971) generally lack a strong geographical component. Comprehensive geographical surveys will only become practicable when the necessary conceptual, methodological and data acquisition issues have been clarified and resolved. At present even the relatively modest existing surveys are fairly complex and expensive, since data have generally had to be collected via household interviews.

A major study of the health care behaviour of 650 households (2,310 individuals) in three suburbs of Adelaide (Stimson and Cleland, 1975; Cleland *et al*., 1977) illustrates this approach. The authors collected extensive data on use patterns of a large number of primary and specialist health care services in 1974. In addition they assessed levels of consumer satisfaction with these services, prior knowledge of services, demographic and socio-economic attributes of patients and households, health insurance status and accessibility to services. Their findings, too substantial to be reviewed adequately here, confirmed the need for an understanding of the nature of consumer behaviour and of how this varies between subgroups of the population. Thus the use patterns of GPs in suburban Adelaide could be explained not only by trip-length differences resulting from variations in local GP:population ratios and practice organisation (that is solo:group practices), but also by social and demographic factors (for example age, sex, family status, socio-economic status, nativity), multi-purpose trips (for example one third of patients visit the GP in association with other activities) and consumer preference for particular GPs. Similar findings have been obtained in other investigations elsewhere in Australia (e.g. Payne *et al.*, 1977), the USA (Pyle, 1979) and the UK (Phillips, 1979, 1981).

The impact of intra-urban variations in primary care availability upon consumer health and behaviour appears to be profound. Indeed Hart (1971) has suggested that an 'inverse law of care' exists, with the availability of good medical care tending to vary inversely with the need of the population served. Several empirical behavioural studies have confirmed that this does occur. The actual and perceived availability and accessibility of medical services evidently have marked effects on spatial patterns of disease, even at the intra-urban scale. Thus high rates of morbidity and mortality for many diseases have been strongly correlated with poor services in both central and certain suburban neighbourhoods in Chicago (for example, de Visé, 1969; Pyle, 1971), Akron (Pyle and Lauer, 1975), Washington, DC (Shannon and Spurlock, 1976), West Glamorgan (Thomas and Phillips, 1978) and Hillsborough County (Todsen, 1980).

These high rates in underprovided areas are generally reinforced by the differential effects of social and behavioural patterns (Joseph, 1979; Fiedler, 1981; Joseph and Boeckh, 1981). Thus sick persons living a long way from medical services will frequently not seek professional help because of the costs (financial, physical and psychological) of travelling to the surgery. For the poor, the aged and many minority groups this behaviour is reinforced by the costs of paying for routine professional treatment. Increasingly, in many countries, these medically indigent groups have come to regard the outpatient and emergency departments of public hospitals as their chief sources of primary and specialist health care (de Visé, 1969, 1973; Kennedy, 1974; Ingram *et al.*, 1978; Thomas and Phillips, 1978; Pyle, 1979; Roghmann and Zastowny, 1979; Phillips, 1981).

For some medical geographers concern with consumer health care behaviour extends beyond the traditional analyses of first contacts with particular elements of the health services system. This is particularly true for mental illness and mental subnormality, which generally require lengthy (often lifelong) treatment and support. In most industrialised countries their patterns of treatment have changed profoundly since the early 1950s, shifting from long term custodial care in large old hospitals to a variety of community based forms of care (for example, psychiatric units in general hospitals, community mental health centres, day patient centres, clinics and welfare agency services). These changes have prompted several medical geographers to analyse the impact of these new facilities and treatment on the behaviour of both consumers (for example, Dear, 1977a, b; Smith, 1978, 1980, 1982) and the residents of neighbourhoods in which new mental health facilities have been proposed or actually built (for example, Boeckh *et al.*, 1980; Dear and Wittman, 1980; Dear *et al.*, 1980; Smith and Hanham, 1981; Taylor *et al.*, 1979).

The Geographer and Health Service Planning

The role of the geographer in health care delivery planning has been quite modest to date, but there is every indication that it could become substantial in the near future. Since the late 1960s the recognition of spatial imperfections in the general efficiency and equity of existing regional and intra-urban health care delivery systems has prompted geographers and other social scientists to offer solutions to these defects. As a result there now exists a voluminous literature dealing with the measurement of the locational efficiency of existing medical care distribution systems and the restructuring of particular elements in those systems.

Many researchers have examined specific elements in the health care delivery system in particular urban settings. Thus in Chicago several workers have focused upon hospitals, the highest tier of the health delivery hierarchy (for example, Morrill and Earickson, 1968, 1969; Morrill and Kelly, 1970). Specialised health services have also been examined in Chicago. Thus Pyle (1971) developed a model to forecast future cancer distributions and the need for cancer teletherapy treatment. The plan was designed spatially to allocate forecasted patients to new installations in 1980. The value of geographical analyses of the primary level of the health care hierarchy is exemplified by the work of Ingram and colleagues in Toronto (1978), Mulvihill in Guatemala City (1979), Curtis (1982) in East Kent and Bennett (1981) in Lansing.

Several geographers have made valuable contributions to what might best be described as 'outcome-orientated' medical geography. The spatial reorganisation of mental health services in North America has been monitored by Wolpert *et al*., (1975), Wolpert and Wolpert (1976) and Smith (1976). Their researches in Canada and the USA suggest that the officially approved policy of decanting the care of the majority of mentally ill persons from large urban mental hospitals into numerous new community centres has proved to be expensive, largely unnecessary and, perhaps, undesirable. The planning and evaluation of optimal emergency medical services delivery is a particularly promising area for geographical work. In the USA the late 1970s witnessed a dramatic increase in expenditure to improve the status of emergency medical services. These services are provided by both private and public agencies at a variety of spatial scales. To date there appear to be few geographically coordinated programmes of emergency health care delivery, yet geographers could play a critical role in their planning and implementation processes. This fact is confirmed by Mayer's (1979a,b, 1981a,b) studies of cardiac arrest cases handled by the Seattle Fire Department Paramedic Teams. These clearly showed that paramedic response times were crucial determinants of long-term survival and that many lives might be saved by locating paramedic vehicles efficiently. Several studies in other localities underline the importance of both the time factor and of planning in the location and operating of emergency medical services (Achabal, 1978; Millner and Goldberg, 1978; Achabal and Schoeman, 1979; Williams and Shavlick, 1979; Monroe, 1980).

An important element in health care planning research (and of planning for other public facilities) has been the development of a large range of models and techniques to optimise facility locations and accessibility. Comprehensive reviews of progress in this field have been provided recently by Veeder (1975), Hodgart (1978), Bach (1980, 1981), Weilbull (1980) and Leonardi (1981a, b). Much of the design and evaluative work now involves the use of computers and computer graphics (for example, Achabal *et al*., 1978; Harner and Slater, 1980).

The role of the medical geographer in health care delivery planning in the future will probably be both enhanced and modified by the requirements of large scale research projects and comprehensive planning. In many countries there has been an increased interest among governments in the regionalisation of health services. In consequence the legislative, functional and spatial frameworks for comprehensive planning have been established. Thus in the USA the passing of the National Health Planning and Resources Development Act in 1974 resulted in the creation of an integrated set of health service agencies and an administratively defined set of Health System Agency Areas

(Pyle, 1979). These areas generally have populations ranging between 500,000 and three millions and most correspond with existing SMSAs. The new Health Systems Agencies (HSAs) have to gather and analyse data relevant for planning purposes, prepare Health Systems Plans (HSP) and Annual Implementation Plans (AIP). Similar developments have also taken place in Australia during the 1970s (Stimson, 1980).

Geographers clearly have many of the skills which are required if these comprehensive schemes are to be established, developed and monitored effectively. Pyle (1979, p. 245) suggests that geographers should collaborate with others involved in health care delivery at the local (that is HSA) level. Here one of the basic prerequisites for planning is the establishment of a geographic base file containing data for small subareas (for example, tracts or EDs). These files have to contain a vast amount of information if the service needs of the population of any particular locality are to be properly assessed and satisfied. In the USA the types of information required have been specified in monographs published by the National Institute of Mental Health (for example, Goldsmith *et al.*, 1975; Rosen *et al.*, 1975; Levy, 1972).

Functioning base files for official census data have been widely available for planning purposes in several countries since the early 1960s. Some are organised on unified computerised (that is automated) mapping and census data retrieval systems. These are best developed in the USA. There the Bureau of Census has created geographic base files (known as DIME — Dual Independent Map Encoding) for over 300 urbanised areas (Pyle, 1979). Similar files and mapping systems exist for specific localities in the UK (Giggs and Mather, 1982) and Australia (Stimson, 1980).

The census data files need to be matched with local health data files. In this way such phenomena as the prevalence and distribution of disease morbidity and mortality, and existing health service utilisation patterns can be identified and mapped. Unfortunately relatively few countries have detailed computerised address-based health files which are continuously updated. There are many such files in the USA, although Pyle (1979, p. 257) observes that they are not yet widely used for research or planning purposes. The development and use of such files has been reported for Adelaide (Stimson, 1980, 1981) and for Nottingham (Giggs, 1982; Giggs and Mather, 1982). Although research in this field is still in a pioneer stage and is hampered by substantial data source problems there is ample evidence to suggest that the development of improved systems of routine reporting of vital statistics and health service data will be the major long-term contribution that geographers will make to those delivering health care.

References

Achabal, D.D. (1978) 'The Development of a Spatial Delivery System for Emergency Medical Services', *Geog. Analysis*, *10 (1)*, 47–64

Achabal, D.D., Moellering, H., Osleeb, J.P. and Swain, R.W. (1978) 'Designing and Evaluating a Health Care Delivery System through the use of Interactive Computer Graphics', *Soc. Sci. Med.*, *12D*, 1–6

Achabal, D.D. and Schoeman, M.E.F. (1979) 'An Examination of Alternative Emergency Ambulance Systems: Contributions from an Economic Geography Perspective', *Soc. Sci & Med.*, *13D*, 81–6

Adesina, H.O. (1981) 'A Statistical Analysis of the Distribution of Characteristics of Cholera within Ibadan City, Nigeria (1971)', *Soc. Sci. Med.*, *15D (1)*, 121–32

Angulo, J.J., Haggett, P. and Pederneiras, C.A.A. (1977) 'Variola Minor in Braganca Paulista County, 1956: A Trend-surface Analysis', *Amer. Journ. Epidem.*, *105*, 272–80

Angulo, J.J., Takiguti, C.K., Pederneiras, C.A.A., Carvalho-de-Souza, A.M., Oliveira-de-Souza, M.C. and Megale, P. (1979) 'Identification of Pattern and Process in the Spread of a Contagious Disease', *Soc. Sci. & Med.*, *13D*, 183–9

Australian Medical Association (1972) *General Practice and its Future in Australia*, Study Group on Medical Planning. A.M.A., Sydney.

Bach, L. (1980) 'Location Models for Systems of Private and Public Facilities based on Concepts of Accessibility and Access Opportunity', *Envt. & Plann.*, *A12*, 301–20

Bach, L. (1981) 'The Problem of Aggregation and Distance for Analyses of Accessibility and Access Opportunity in Location – Allocation Models', *Envt. & Plann.*, *A*, *13 (8)*, 955–78

Baczkowski, D.M. (1980) 'Viral Hepatitis in Metropolitan Sydney', *Australian Geogr.*, *14 (5)*, 285–95

Bagley, C. and Jacobson, S. (1976) 'Ecological Variation of Three Types of Suicide', *Psychol. Med.*, *6*, 423–7

Barnett, J.R. (1978) 'Race and Physician Location: Trends in Two New Zealand Urban Areas', *N.Z. Geogr.*, *34 (1)*, 2–12

Bashshur, R.L., Shannon, G.W. and Metzner, C.A. (1970) 'The Application of Three-dimensional Analogue Models to the Distribution of Medical Care Facilities', *Med. Care*, *8*, 395–407

Bennett, W.D. (1981) 'A Locational-allocation Approach to Health Care Facility Location: A Study of the Undoctored Population in Lansing, Michigan', *Soc. Sci. Med.*, *15D (2)*, 305–12

Boeckh, J., Dear, M. and Taylor, S.M. (1980) 'Property Values and Mental Health Facilities in Metropolitan Toronto', *The Canad. Geogr.*, *24 (3)*, 270–85

Bohland, J.R. and Frech, P. (1982) 'Spatial Aspects of Primary Health Care for the Elderly' in A.M. Warnes (ed.) *Geographical Perspectives on the Elderly*, J. Wiley & Sons, New York, 339–54

Briggs, R. and Leonard, W.A. (1977) 'Mortality and Ecological Structure: A Canonical Approach', *Soc. Sci. & Med.*, *11*, 757–62

Brownlea, A.A. (1967) 'An Urban Ecology of Infectious Disease', *Aust. Geogr.*, *10*, 169-88

Brownlea, A.A. (1972) 'Modelling the Geographic Epidemiology of Infectious Hepatitis' in N.D. McGlashan (ed.) *Medical Geography: Techniques and Field Studies*, London, Methuen & Co. Ltd., 279-300

Burnley, I.H. (1977) 'Mortality Variations in an Australian Metropolis' in N.D. McGlashan (ed.) *Studies in Australian Mortality*, Environmental studies, Occasional Paper no. 4, University of Tasmania, Tasmania, 29-61

Cleland, E.A., Stimson, R.J. and Goldsworthy, A.J. (1977) *Suburban Health Care Behaviour in Adelaide*, Centre for Applied Social and Survey Research, Monograph Series 2. Flinders University, Adelaide

Curtis, S.E. (1982) 'Spatial Analysis of Surgery Locations in General Practice', *Soc. Sci. & Med.*, *16(3)*, 303-13

Dear, M. (1977a) 'Locational Factors in the Demand for Mental Health Care', *Econ. Geogr.*, *53*, 223-40

Dear, M. (1977b) 'Psychiatric Patients and the Inner City', *Ann. Ass. Amer. Geogr.*, *67*, 588-94

Dear, M.J. and Wittman, I. (1980) 'Conflict over the location of mental health facilities' in D.T. Herbert and R.J. Johnston (eds.) *Geography and the Urban Environment: Progress in Research and Applications*, vol. 3, John Wiley & Sons, Chichester

Dear, M., Taylor, S.M. and Hall, G.B. (1980) 'External Effects of Mental Health Facilities', *Ann. Ass. Amer. Geogr.*, *70*, 342-52

de Visé, P. (1969) *Slum Medicine: Chicago's Apartheid Health System*, Community and Family Study Center, Report no. 6, University of Chicago, Chicago, Ill

de Visé, P. (1973) 'Misused and Misplaced Hospitals and Doctors: A Locational Analysis of the Urban Health Care Crisis', *Ass. Amer. Geogrs.*, CCG Resource Paper no. 22, Washington, DC

Dever, G.E.A. (1972) 'Leukaemia and Housing: An Intra-urban Analysis' in N.D. McGlashan (ed.) *Medical Geography: Techniques and Field Studies*, Methuen, London, 233-45

Dewey, D. (1973) 'Where the Doctors have gone: the Changing Distribution of Private Practice Physicians in the Chicago Metropolitan Area, 1950-1970', Research paper, Illinois Regional Medical Program, Chicago Regional Hospital Study, Chicago, Ill

Dickinson, F.G. (1954) *Distribution of Physicians by Medical Service Areas*, Bulletin 94, Bureau of Medical Economic Research, American Medical Association, Chicago

Diseker, R.A. and Chappell, J.A. (1976) 'Relative Importance of Variables in Determination of Practice Location: A Pilot Study', *Soc. Sci. & Med.*, *10*, 559-63

Dorsey, J.L. (1969) 'Physician Distribution in Boston and Brookline: 1940 and 1961', *Medical Care*, *7(6)*, 429-40

Earickson, R. (1970) *The Spatial Behaviour of Hospital Patients: A Behavioural Approach to Spatial Interaction in Metropolitan Chicago*, Research Paper no. 124, Dept. of Geography, University of Chicago, Chicago

Eisenberg, B.S. and Cantwell, J.R. (1976) 'Policies to Influence the Spatial Distribution of Physicians: A Conceptual Review of Selected Programs and Empirical Evidence', *Med. Care*, *14(6)*, 455-68

Elesh, D. and Schollaert, P.T. (1972) 'Race and Urban Medicine: Factors Affecting the Distribution of Physicians in Chicago', *Journ. Health & Soc. Behav.*, *13*, 236-50

Fiedler, J.L. (1981) 'A Review of the Literature on Access and Utilization of

Medical Care with Special Emphasis on Rural Primary Care', *Soc. Sci. & Med.*, *15C(3)*, 129–42

Florin, J.W. (1971) *Death in New England: Regional Variations in Mortality*, Chapel Hill, University of North Carolina, Dept. of Geography

Ford, A.B. (1976) *Urban Health in America*, Oxford University Press, Oxford

Freestone, R. (1975) 'On Urban Resource Allocation: the Distribution of Medical Practitioners in Sydney', *Geog. Bull. NSW Geog. Soc.*, pp. 14–25

Giggs, J.A. (1979) 'Human Health Problems in Urban Areas' in D.T. Herbert and D.M. Smith (eds.) *Social Problems and the City: Geographical Perspectives*, Oxford University Press, London, pp. 84–116

Giggs, J.A. (1980) 'Mental Health and the Environment' in G.M. Howe and J.A. Loraine (eds.) *Environmental Medicine*, Heinemann Medical Books Ltd., London, pp. 281–306

Giggs, J.A., Ebdon, D.S. and Bourke, J.B. (1980) 'The Epidemiology of Primary Acute Pancreatitis in the Nottingham Defined Population Area', *Trans., Inst. Brit. Geogr.*, *5(2)*, 229–42

Giggs, J.A. (1982) 'Schizophrenia and Ecological Structure in Nottingham' in N. McGlashan (ed.) *Space and Health*, Academic Press, London

Giggs, J.A. and Mather, P.M. (1982) *Social Ecology and Mental Health in Nottingham*, Report Series in Applied Geography, no. 2, Dept. of Geography, University of Nottingham

Gilbert, E.W. (1958) 'Pioneer Maps of Health and Disease in England', *Geog. Journal*, *124*, 172–83

Gilbert, E.W. (1972) *British Pioneers in Geography*, David & Charles, Newton Abbot

Girt, J.L. (1972) 'Simple Chronic Bronchitis and Urban Ecological Structure' in N.D. McGlashan (ed.) *Medical Geography: Techniques and Field Studies*, Methuen, London, 219–31

Gober, P. and Gordon, R.J. (1980) 'Intra-urban Physician Location: A Case Study of Phoenix', *Soc. Sci. & Med.*, *14D*, 407–17

Goldsmith, H.F. *et al.*, (1975) '*A Typological Approach to doing Social Area Analysis*', NIMH, DHEW Publications no. (ADM) 76–262 US Govt. Printing Office, Washington DC

Gross, P.F. (1974) (ed.) 'Technology, Society and Health Care', *Search*, *5(10)*, Australian and New Zealand Association for the Advancement of Science

Guptill, S.C. (1975) 'The Spatial Availability of Physicians', *Proc. Assoc. Amer. Geogr.*, *7*, 80–4

Guzick, D.S. and Jahiel, R.I. (1976) 'Distribution of Private Practice Offices of Physicians with Specified Characteristics among Urban Neighbourhoods', *Med. Care*, *14(6)*, 469–58

Haggett, P. (1976) 'Hybridizing Alternative Models of an Epidemic Diffusion Process', *Econ. Geogr.*, *52*, 136–46

Harner, E.J. and Slater, P.B. (1980) 'Identifying Medical Regions Using Hierarchical Clustering', *Soc. Sci. Med.*, *14D*, 3–10

Harrison, G.A. and Gibson, J.B. (1976) (eds.) *Man in Urban Environments*, Oxford University Press, London

Hart, J.T. (1971) 'The Inverse Care Law', *Lancet*, 405–12. Also in C. Cox and A. Mead (eds.) *A Sociology of Medical Practice*, Collier-MacMillan, London, 189–206

Herbert, D.T. (1976) 'Social Deviance in the City: A Spatial Perspective' in D.T. Herbert and R.J. Johnston (eds.) *Social Areas in Cities*, vol. 2, John Wiley & Sons, London, 89–121

Hirsch, A. (1883–6) *Handbook of Geographical and Historical Pathology*, 3 vols. (Trans. C. Creighten, Jr) New Sydenham Society, London

Hodgart, R.L. (1978) 'Optimising Access to Public Services: A Review of the Problems, Models and Methods of Locating Central Facilities', *Prog. in Hum. Geogr.*, *2*, 17-48

Howe, G.M. (1979) 'Death in London', *Geog. Mag.*, *Li*, *4*, 284-9

Ingram, D.R., Clarke, D.R. and Murdie, R.A. (1978) 'Distance and the Decision to Visit an Emergency Department', *Soc. Sci. Med.*, *12D*, 55-62

Jacobson, S., Bagley, C. and Rehin, A. (1976) 'Completed Suicide: A Taxonomic Analysis of Clinical and Social Data', *Psychol. Med.*, *6*, 429-38

Joseph, A.E. (1979) 'The Referral System as a Modifier of Distance Decay Effects in the Utilization of Mental Health Care Services', *The Canad. Geogr.*, *23(2)*, 159-69

Joseph, A.E. and Boeckh, J.L. (1981) 'Locational Variation in Mental Health Care Utilization Dependent upon Diagnosis: A Canadian Example', *Soc. Sci. & Med.*, *15D(3)*, 395-404

Kaplan, B.H. (1971) (ed.) *Psychiatric Disorder and the Urban Environment*, Behavioural Publications, New York

Kasl, S.V. (1977) 'The Effects of the Residential Environment on Health and Behaviour' in L.E. Hinkle and W.C. Loring (eds.) *The Effects of the Man-made Environment on Health and Behaviour*, DHEW Publications, no. (CDC) 77-8318, Washington, DC, 65-128

Keig, G. and McAlpine, J.R. (1980) 'The Influence of Age in Analysis of Mortality Variation between Population Groups', *Soc. Sci. & Med.*, *14D*, 165-8

Kennedy, N. (1974) 'Community Use of Out-patient and Casualty Services at the Adelaide Children's Hospital' in E.A. Cleland and R.J. Stimson (eds.) *The Delivery and Evaluation of Health Care Services: Proceedings of a Symposium*, Flinders Health Care Services Evaluation Research Group, Pubn., no. 1, Flinders University, Adelaide

Klauber, M.R. and Angulo, J.J. (1976) 'Three Tests for Randomness of Attack of Social Groups during an Epidemic', *Amer. Journ. Epidem.*, *104*, 212-8

Knox, G. (1964) 'Detection of Time-space Interactions', *Appl. Statist.*, *13*, 25-9

Knox, G. (1971) 'Epidemics of Rare Disease', *Brit. Med. Bull.*, *27*, 43-7

Knox, P. (1978) 'The Intra-urban Ecology of Primary Medical Care: Patterns of Accessibility and their Policy Implications', *Envt. & Plann.*, *A10*, 425-35

Knox, P. (1979) 'The Accessibility of Primary Care to Urban Patients: A Geographical Analysis', *J. Royal Coll. Gen. Practit.*, *29*, 160-8

Knox, P. and Pacione, M. (1980) 'Locational Behaviour, Place Preferences and the Inverse Care Law in the Distribution of Primary Medical Care', *Geoforum*, *11*, 43-55

Krupinski, J. and Stoller, A. (1971) (eds.) *The Health of a Metropolis*, Heinemann Educational Books, Melbourne

Lankford, P.M. (1974) 'Physician Location Factors and Public Policy', *Econ. Geog.*, *50*, 244-55

Learmonth, A. (1975) 'Ecological Medical Geography', *Progress in Geography*, 201-26

Leonardi, G. (1981a) 'A Unifying Framework for Public Facility Location Problems – part 1: A Critical Overview and some Unresolved Problems', *Envt. & Plann. A.*, *13(8)*, 1001-28

Leonardi, G. (1981b) 'A Unifying Model for Public Facility Location Problems – part 2: some New Models and Extensions', *Envt. & Plann. A*, *13(9)*, 1085-108

Levy, R.M. (1972) *Use of a Geographic Base File in a Health Information System*, Computerized Geographic Coding Series GE 60 no. 3, US Dept. of Commerce, Bureau of the Census, Washington, DC

McGlashan, N.D. (1972) (ed.) *Medical Geography: Techniques and Field Studies*, Methuen, London

McGlashan, N.D. (1977) 'A Note on the Medical Case Hierarchy', *Soc. Sci. & Med.*, *11*, 773–4

Martin, A.E. (1964) 'Mortality and Morbidity Statistics and Air Pollution', *Proc. Roy. Soc. Med.*, *57*, 969–74

Mayer, J.D. (1979a) 'Seattle's Paramedic Program: Geographical Distribution, Response Times and Mortality', *Soc. Sci. & Med.*, *13D*, 45–51

Mayer, J.D. (1979b) 'Paramedic Response Time and Survival from Cardiac Arrest', *Soc. Sci. & Med.*, *13D*, 267–71

Mayer, J.D. (1981a) 'Geographic Patterns of Cardiac Arrest: An Exploratory Model', *Soc. Sci. & Med.*, *15D (3)*, 329–34

Mayer, J.D. (1981b) 'A Method for the Geographical Evaluation of Emergency Medical Services Performance', *Amer. J. Publ. Health*, *71 (8)*, 841–4

Miller, A.E. (1977) 'The Changing Structure of the Medical Profession in Urban and Suburban Settings', *Soc. Sci. & Med.*, *11*, 233–43

Miller, D.S. *et al.*, (1975) 'Cronin's Disease in Nottingham: A Search for Time-space Clustering', *Gut*, *16*, 454–7

Millner, R. and Goldberg, J. (1978) 'Toward an Outcome-oriented Medical Geography: An Evaluation of the Illinois Trauma/Emergency Medical Services System', *Soc. Sci. & Med.*, *12D*, 103–10

Ministry of Health (1954) *Mortality and Morbidity during the London Fog of December 1952*, HMSO, London

Monroe, C.B. (1980) 'A Simulation Model for Planning Emergency Response Systems', *Soc. Sci. Med.*, *14D*, 71–7

Moos, R.R. (1976) *The Human Context: Environmental Determinants of Behavior*, John Wiley & Sons, New York

Morrill, R.L. (1959) 'Empirical studies of physician utilization' in W.L. Garrison *et al.*, (eds.) *Studies of Highway Development and Geographic Change*, University of Washington Press, Seattle

Morrill, R.L. (1966) *Historical Development of the Chicago Hospital System*, Chicago Regional Hospital working paper no. 1, 2, Hospital Planning Council of Chicago, Chicago

Morrill, R.L. and Angulo, J.J. (1979) 'Spatial Aspects of a Smallpox Epidemic in a Small Brazilian City', *Geog. Rev.*, *69*, 319–30

Morrill, R.L. and Angulo, J.J. (1981) 'Multivariate Analysis of the Role of School Attendance Status in the Introduction of Variola Minor into the Household', *Soc. Sci. & Med.*, *15D (4)*, 479–87

Morrill, R.L. and Earickson, R. (1968) 'Variations in the Character and Use of Hospital Services', *Health Services Research* 3, reprinted in L.S. Bourne (ed.) *Internal Structure of the City*, 1971, (London) Oxford University Press, 391–9

Morrill, R.L. and Earickson, R.J. (1969) 'Locational Efficiency of Chicago Hospitals: An Experimental Model', *Hlth. Serv. Res.*, *4*, 128–41

Morrill, R.L. Earickson, R.J. and Rees, P. (1970) 'Factors Influencing Distances Travelled to Hospitals', *Econ. Geogr.*, *46*, 161–71

Morrill, R.L. and Kelley, M.B. (1970) 'The Simulation of Hospital Use and the Estimation of Location Efficiency', *Geogr. Anal.*, *2*, 283–300

Morris, J. (1976) 'Access to Community Health Facilities in Melbourne' in Australian and New Zealand Section, Regional Science Association and Department of Geography, University of Queensland, Brisbane, 143–73

Moyes, A. (1977) 'Accessibility to General Practitioner Services on Anglesey: some Trip-making Implications', Paper presented to the Institute of British Geographers, Newcastle-upon-Tyne

Mulvihill, J.L. (1979) 'A Location Study of Primary Health Services in Guatemala

City', *Prof. Geogr.*, *31 (3)*, 299–305

Norcliffe, G.B. (1977) *Inferential Statistics for Geographers*, Hutchinson University Library, London

Offices of Health Economics (1974) *The work of primary medical care*, no. 49 in series on current health problems, OHE, London

Patno, M.E. (1954) 'A Geographical Study of Cancer Prevalence Within an Urban Population', *Public Health Reports*, *69 (8)*, 705–15

Payne, S. *et al.*, (1977) 'Accessibility factors in health care utilization', ANZSERCH Proceedings, Adelaide, 18–28

Phillips, D.R. (1979) 'Spatial Variations in Attendance at General Practitioner Services', *Soc. Sci. & Med.*, *13D*, 169–81

Phillips, D.R. (1981) *Contemporary Issues in the Geography of Health Care*, Norwich, Geo Books

Pyle, G.F. (1971) 'Heart Disease, Cancer and Stroke in Chicago', Dept. of Geography Research paper no. 124, University of Chicago, Chicago

Pyle, G.F. (1973) 'Measles as an Urban Health Problem: the Akron Example', *Econ. Geogr.*, *49*, 344–56

Pyle, G.F. (1977) 'International Communication and Medical Geography', *Soc. Sci. & Med.*, *11*, 679–82

Pyle, G.F. (1979) *Applied Medical Geography*, V.H. Winston & Sons, Washington DC

Pyle, G.F. and Lauer, B.M. (1975) 'Comparing Spatial Configuration: Hospital Service Areas and Disease Rates', *Econ. Geogr.*, *51*, 50–68

Robertson, L.S. (1970 'On the Intra-urban Ecology of Primary Care Physicians', *Soc. Sci. & Med.*, *4*, 227–38

Roemer, M.I. (1977) *Systems of Health Care*, Springer, New York

Roghmann, K.J. and Zastowny, T.R. (1979) 'Proximity as a Factor in the Selection of Health Care Providers: Emergency Room Visits Compared to Obstetric Admissions and Abortions', *Soc. Sci. & Med.*, *13D*, 61–9

Rosen, B.M. *et al.*, (1975) *Mental Health Demographic Profile System Description: Purpose, Contents and Sampler of Uses*, NIMH, DHEW Publications no. (ADM) 76–263, US Govt. Printing Office, Washington, DC

Rosenthal, S.F. (1978) 'Target Populations and Physician Populations: the effects of Density and Change', *Soc. Sci. & Med.*, *12D*, 111–15

Royal College of Physicians of London (1970) *Air Pollution and Health*, Pitmans, London

Sainsbury, P. (1955) *Suicide in London*, Maudsley Monography no. 1, Chapman and Hall, London

Schneider, J.B. (1967) 'Measuring the Locational Efficiency of the Urban Hospital', *Hlth. Serv. Res.*, *2*, 154–69

Shannon, G.W. (1977) 'Space, Time and Illness Behaviour', *Soc. Sci. & Med.*, *11*, 683–9

Shannon, G.W. and Dever, G.E.A. (1974) *Health Care Delivery: Spatial Perspectives*, McGraw-Hill, New York

Shannon, G.W. and Spurlock, C.W. (1976) 'Urban Ecological Containers, Environmental Risk Calls, and the use of Medical Services', *Econ. Geogr.*, *52*, 171–80

Shannon, G.W., Bashshur, R.L. nd Metzner, C.A. (1969) 'The Concept of Distance as a Factor in Accessibility and Utilization of Health Care', *Med. Care Review*, *26*, 143–61

Shannon, G.W., Skinner, J.L. and Bashshur, R.L. (1973) 'Time and Distance: the Journey for Medical Care', *Int. J. Hlth. Serv.*, *3(2)*, 237–43

Shannon, G.W., Spurlock, C.W. and Skinner, J.L. (1975) 'A Method for Evaluating the Geographic Accessibility of Services', *Prof. Geog.*, *27(1)*,

30–6

Smith, C.J. (1976) 'Distance and the Location of Community Mental Health Facilities: A Divergent Viewpoint', *Econ. Geogr.*, *52*, 181–92

Smith, C.J. (1977) *The Geography of Mental Health*, Resource paper no. 76–4, Association of American Geographers, Washington, DC

Smith, C.J. (1978) 'Recidivism and Community Adjustment amongst Former Mental Patients', *Soc. Sci. & Med.*, *12D*, 17–27

Smith, C.J. (1980) 'Neighbourhood effects on mental health' in D.T. Herbert and R.J. Johnston (eds.) *Geography and the Urban environment*, vol. 3, John Wiley & Sons, Chichester, 363–416

Smith, C.J. (1982) 'Home-based mental health care for the elderly' in A.M. Warnes (ed.) *Geographical Perspectives on the Elderly*, John Wiley & Sons, New York, 375–98

Smith, C.J. and Hanham, R.Q. (1981) 'Any Place but Here! Mental Health Facilities as Noxious Neighbors', *Prof. Geogr.*, *33(3)*, 326–34

Srole, L. (1972) 'Urbanisation and Mental Health: some Reformulations', *Amer. Scientist*, *60*, 576–83

Stimson, R.J. (1980) 'Spatial Aspects of Epidemiological Phenomena and of the Provision and Utilization of Health Care Services in Australia: A Review of Methodological Problems and Empirical Analyses', *Envt. & Plann.*, *A. 12*, 881–907

Stimson, R.J. (1981) 'The Provision and use of General Practitioner Services in Adelaide, Australia: Application of Tools of Locational Analysis and Theories of Provider and User Spatial Behaviour', *Soc. Sci. & Med.*, *15D(1)*, 27–44

Stimson, R.J. and Cleland, E.A. (1975) 'Household Health Care Behaviour in Suburban Areas of Adelaide: A Baseline Study' in J.S. Dodge and S.R. West (eds.) *Epidemiology and Primary Medical Care*, Dunedin, 84–91

Taylor, S.M., Dear, M.J. and Hall, G.B. (1979) 'Attitudes toward the Mentally ill and Reactions to Mental Health Facilities', *Soc. Sci. & Med.*, *13D*, 281–90

Thomas, C.J. (1976) 'Sociospatial Differentiation and the use of Services' in D.T. Herbert and R.J. Johnston (eds.) *Social Areas in Cities*, vol. 2, John Wiley & Sons, London 17–63

Thomas, C.T. and Phillips, D.R. (1978) 'An Ecological Analysis of Child Medical Emergency Admissions to Hospitals in West Glamorgan', *Soc. Sci. & Med.*, *12D*, 183–92

Todsen, D.R. (1980) 'Spatial Perspectives of Infant Health Care: the Distribution of Infant Health Care Delivery in Hillsborough County, Florida', *Soc. Sci. & Med.*, *14D*, 379–85

Veeder, N.W. (1975) 'Health Service Utilization Models for Human Services Planning', *J. Inst. Amer. Plann.*, *41*, 101–09

Webber, R.J. (1975) *Liverpool Social Area Study, 1971 Data: Final Report*, PRAG Technical Paper TP14, Centre for Environmental Studies, London

Weibull, J.W. (1980) 'On the Numerical Measurement of Accessibility', *Envt. & Plann.*, *12*, 53–67

Weinstein, M.S. (1980) *Health in the City*, Pergamon Press, New York

Whitlock, F.A. (1973) 'Suicide in England and Wales 1959–63: Part 2: London', *Psychol. Med.*, *3*, 411–20

Wilkins, E.T. (1954) 'Air Pollution Aspects of the London Fog of December 1952', *Q.J.R. Met. Soc.*, *80*, 267–71

Williams, P.M. and Shavlick, G. (1979) 'Geographic Patterns and Demographic Correlates of Paramedic Runs in San Bernadino, 1977', *Soc. Sci. & Med.*, *13D*, 273–80

Wilson, M.G.A. (1978) 'The Geographical Analysis of Small area/population Death Rates – A Methodological Problem', *Austral. Geogr. Stud.*, *16(Z)*,

149-60

Wolpert, J., Dear, M. and Crawford, R. (1975) 'Satellite Mental Health Facilities', *Ann. Ass. Amer. Geogr.*, *65*, 24-35

Wolpert, F. and Wolpert, E. (1976) 'The Relocation of Released Mental Hospital Patients into Residential Communities', *Policy Science*, 7, 31-51

Wood, C.M., Lees, N., Luker, J.A. and Saunders, P.J.W. (1974) *The Geography of Pollution: A Study of Greater Manchester*, Manchester University Press, Manchester

Wood, C.M. and Lawrence, M. (1980) 'Air Pollution and Human Health in Greater Manchester', *Envt. & Plann.*, *A. 12*, 1427-39

9 TERRITORIAL JUSTICE AND SERVICE ALLOCATION

A.M. Kirby and S.P. Pinch

One of the most visible changes in urban geography in the last decade has been the enormous upsurge of research into service provision and public facility location within cities. The range of services studied has been remarkable including public-sector housing (Pinch, 1978), housing improvement grants (Duncan, 1974), housing finance (Bassett and Short, 1980), housing inspections (Nivola, 1979), rat control (Margolis, 1977), sanitation inspection (Jones *et al.*, 1980), doctors surgeries (Knox, 1978), dental facilities (Bradley, Kirby and Taylor, 1978), social services for the elderly (Pinch, 1979), hospitals (Morrill *et al.*, 1970), mental health facilities (Dear, 1977), nurseries (Freestone, 1977), schools (Kirby, 1978), parks (Mladenka and Hill, 1977), libraries (Levy *et al.*, 1974), fire services (Seley, 1979), police calls (Mladenka and Hill, 1978), planning permissions (Simmie, 1978) and many others (for reviews see Bennett, 1980; Burnett, 1981; Jones *et al.*, 1980; Rich, 1979; Webster, 1981).

These studies have examined patterns at numerous geographical scales — both between and within regions, cities, and various types of political and administrative unit. Furthermore, there has been a welter of methodological perspectives ranging from liberal-democratic to structuralist-Marxist. The field is truly inter-disciplinary in character attracting political scientists, sociologists, economists and social administrators, as well as geographers (who have often played a leading role) but is united by a common interest in the distributional features of service provision.

This diversity is undoubtedly a healthy indication of a dynamic and evolving subject matter but means that the field is often impenetrable for both newcomer and established researcher alike. It is therefore essential that there are regular examinations of progress and this chapter is intended as one such stock-taking. The first part discusses the origins of geographical studies of service provision and examines the major conclusions to have emerged from empirical research. The rest of the chapter examines three major theoretical perspectives that are most frequently offered to account for variations in service provision — pluralism, managerialism, and structuralism.

223

The Origins and Form of Spatial Studies of Service Provision

The origins of the recent interest in service provision may be traced back to a series of studies undertaken by North American economists which attempted to explain variations in local government expenditure using multiple regression (Colm, 1936; Brazer, 1959; Fisher, 1961; Sacks and Harris, 1964). Their principal finding was that, unlike environmental factors such as population density, degree of urbanisation or *per capita* incomes, political variables had little or no effect upon expenditures. These results were generally corroborated by a series of studies undertaken by political scientists (for example, Dawson and Robinson, 1963; Dye, 1966; Hofferbert, 1966) but enormous controversy and a vast literature ensued (for reviews see Fenton and Chamberlayne, 1969; Sharkansky and Hofferbert, 1969). This type of analysis eventually developed in Britain (for example, Alt, 1971; Boaden, 1971; Nicholson and Topham, 1971) where generally political variables appear to have greater explanatory power (for reviews and criticisms see Alt, 1977; Le Grand and Winter, 1977; Newton and Sharpe, 1977).

The origins of the recent geographical interest in service provision *within* cities came later in the early seventies. Of particular importance at this time was Harvey's (1971) seminal discussion of the relationship between social process and the spatial form of the city. Harvey drew attention to the way in which, as we change the spatial form of the city, by relocating employment opportunities, housing, transport routes, sources of pollution and the like, so we change two factors: first the *price of accessibility* to those urban facilities which are either desired or necessary, and second, the *costs of proximity* to those facilities which are considered undesirable. The net effect of these positive and negative externalities, filtered through individual perceptions and social value systems, Harvey termed 'real income'. The inspiration for this concept was derived from the view of Titmuss that:

> No concept of income can be really equitable that stops short of the comprehensive definition which embraces all receipts which increase an individual's command over the use of a society's scarce resources . . . (Titmuss in Harvey, 1973, p. 53).

The crucial point was that the distribution of real income may, to some degree, be independent of, but at the same time may reinforce inequalities of direct monetary income derived from the occupational market.

Much of Harvey's analysis was intuitive and speculative in character, but he argued, on the basis of certain assumptions, that the more affluent would organise themselves through the political system to maximise their real income, thus leading to considerable inequality in resource allocation. Harvey then proceeded to consider which principles ought to govern the allocation of resources in cities. Following from the earlier work of Davies (1968), he argued that 'spatial' or 'territorial' justice would ensue from the allocation of resources on the basis of 'need' (Harvey, 1973).

A similar argument was developed at the same time by Pahl (1970). He stressed that there were two fundamental constraints upon individuals' access to scarce resources in urban areas. The first he termed *spatial* and might be expressed in terms of the time or cost involved in overcoming the friction of distance when travelling to a facility. The second type of constraint he regarded as *social*, which would reflect the distribution of power in society. In the second case access to scarce resources was governed by bureaucratic rules and procedures of the local managers in the urban system whom he labelled 'social gatekeepers'.

As in the case of Harvey's paper, many of Pahl's ideas were speculative and unrefined and it is not surprising that they were later subject to considerable amendment. Pahl's work spawned the controversial 'managerialist' approach which he soon rejected in favour of a form of Corporatism (Pahl, 1977). Harvey, appreciating the operational difficulties embodied in his all embracing notion of 'real income', rapidly rejected this liberal approach. He then proceeded within a neo-Marxist framework to link notions of under-consumption to the growth of suburbs in Western capitalist societies (Harvey, 1977). Nevertheless, both Harvey and Pahl's work encouraged the belief that 'local public services bid fair to become the chief means of income distribution in our economy' (Thompson, 1965) and thus stimulated their analysis.

Two basic questions have been asked in the following decade. The first is concerned with the differential allocation of services between individuals or subgroups, or put more simply in Lasswell's (1958) now famous terminology 'who gets what?'. This question has obvious geographical implications for services are located in, or delivered to, particular areas or neighbourhoods — hence the geographical interest in 'who gets what *where*?' (Smith, 1977). The second basic question is concerned with uncovering the mechanisms responsible for these patterns of service allocation or again in Lasswell's terms 'how?' (to which can also be added the question 'why?').

Although many complex taxonomic schemes have been applied to service provision, at a basic level two types of service can be distinguished. First, there are the relatively 'pure' public goods such as parks, libraries, swimming pools, museums, police and fire protection which are theoretically available to all members of society. Second there are the highly 'impure' services such as education, public housing, and welfare services that are restricted to particular subgroups or minorities. These are sometimes called 'merit' goods (see Margolis, 1968).

The original formulations of Samuelson (1954) and Musgrove (1959) defined the first type of pure public goods in idealised terms by reference to three main criteria. First there is the notion of *joint supply*, which implies that if the good is supplied to one person it can be supplied to all others at no extra cost. Second, is the criteria of *non-excludability* whereby having supplied the good to one person it is impossible to withhold the good from others so that those who do not pay for the good cannot be prevented from its benefits. Third is the idea of *non-rejectability*, which means that once a good or service is supplied it must be equally consumed by all even if some do not wish to do so. These criteria define pure public goods as the polar opposite of private goods but in practice most public goods and services (including the classic example of defence) will possess varying degrees of 'impurity' (Bennett, 1980).

The factors which most commonly undermine this theoretical impurity are geographical in nature. Most nations are, for administrative reasons, subdivided into numerous smaller local government units or political jurisdictions. For a variety of economic, social, political and administrative reasons these units vary in the extent to which they provide goods and services as demonstrated by the 'expenditure determinant' studies. Within these jurisdictions the location of public facilities also affects their consumption. There is now considerable evidence that individuals are more likely to use nearby facilities and conversely, that the speed with which, for example, emergency services can reach the home will depend upon distance from the service base.

In the second category of 'merit' goods however, the services often have (in principle at least) redistributive welfare aims. Thus socially defined criteria of 'need' become the principle factors which determine who can obtain access to the service. This implies the second type of social constraint as originally defined by Pahl. Thus it is in the realm of welfare services such as child care, social services for the elderly and public housing that most of the effort has been made to evaluate the extent of territorial justice — the relationships between needs and resources.

This distinction between service type and explanatory influence cannot be maintained too rigorously. On the one hand, while public utilities may be intended for all, there may be sound reasons for locating them in certain districts so that certain subpopulations in need have a greater probability of consuming the service. Thus one might add additional library facilities in areas where the inhabitants lacked income to purchase books, or parks in areas of high density dwellings with small or non-existent gardens. On the other hand location cannot be entirely discounted as a factor affecting the consumption of welfare-orientated services. Access to elderly day-care centres may be restricted by receptionists acting in a 'gatekeeper' role but the location of the facilities may also affect the utilisation of the service. Nevertheless, the distinction has affected the orientation of researchers. Those concerned with relatively pure services have developed the theory of 'public facility location' concentrating upon 'spatial' factors by seeking the optimal location in which the aggregate distance between the facility and the surrounding population is minimised (Hodgart, 1978). Those concerned with the welfare orientated services have tended to draw upon a variety of broader social theories to explain patterns of resource allocation. Nevertheless, it should be clear that in the future there will need to be a closer interaction between these two styles of research.

Who gets what where?

An underlying assumption in much of the writing on service provision is some extension of what Lineberry (1977a, 1977b), writing in a North American context, termed 'the underclass hypothesis'. This has been variously expressed as 'the poor pay more', 'them that has gets' and 'unto him who hath shall be given'. These ideas suggest in Sharrard's (1968) now classic words that:

> The slum is the catch-all for the losers, and in the competitive struggle for the cities goods and services the slum areas are also the losers in terms of schools, jobs, garbage collection, street lighting, libraries, social services, and whatever else is communally available but always in short supply (Sharrard, 1968, p. 10).

In the British context the most common expression of the underclass hypothesis has been embodied in the so-called 'inverse-care law'. This was first propounded in a medical context by Tudor-Hart (1971) who claimed that 'the availability of good medical care tends to vary inversely with the need for it in the population'. Stated more simply this means that the greater your need for medical care the less likely you are to receive it. If this process of inverse-care is as all pervasive as the status of a 'law' would suggest, then one might expect to find in the literature on service provision consistent evidence of territorial injustice – an inverse relationship between needs and resources. However, it would appear that such evidence is not available for the geographical patterns are complex. Some studies would seem to support the idea of inverse care. Duncan (1974) examined the allocation of improvement grants to enumeration districts in Huddersfield. He compared the actual number of grants received in each area with an expected share for each area based on the assumption that each enumeration district received a total of improvement grants in proportion to its number of dwellings without the three basic amenities. Improvement grants were seen to make an important contribution in providing amenities for low value housing in the city. However, the concentrations of grant allocations tended to avoid areas with high proportions of rented properties and coloured immigrants – the areas which also had the highest levels of census indices of housing disadvantage.

Knox (1978) also drew inspiration from the notion of inverse-care for his study of the geographical distribution of doctors surgeries in four Scottish cities – Aberdeen, Dundee, Glasgow, and Edinburgh. Inertia was one of the key factors which appeared to affect the distribution of facilities. Surgeries tended to cluster in the older established residential areas of Scottish cities and were noticably absent from the modern peripheral housing estates in both the private and public sectors. However, there was no systematic support for the underclass hypothesis. Although some of the low class (and by inference high need) areas on the outskirts of cities were 'under-provided' there were in some cases concentrations of surgeries near relatively deprived inner-city areas.

Research by Pinch (1978, 1979) on housing and social service allocation for the elderly in greater London indicated a complex set of relationships. One of the most striking features was the way in which the modern community-based home-helps and meals-on-wheels services were more strongly correlated with needs than the more traditional forms of institutionalised care for the elderly in residential accommodation. However, the community-based health services, home nursing and health visiting, were either poorly or negatively correlated with the needs index (Pinch, 1979, 1980). In the field of housing there was relatively little difference between the net construction rate in the relatively low need suburban authorities and the relatively high-need inner boroughs (Pinch, 1978). As in the case of the social services there was little evidence of territorial justice but at the same time little systematic evidence of inverse-care.

A similar diversity of patterns has emerged from the North American context. One of the earliest studies examined services within Oakland (Levy *et al.*, 1974) and found that educational resources were U shaped in character, with relatively higher levels of provision in the poor and rich areas. Expenditure on libraries in contrast went to those areas with the greatest demand — the relatively affluent districts — and neglected the poorer areas with low circulation levels. Roads tended to benefit suburban dwellers.

Lineberry (1977a, 1977b) concluded from his study of San Antonio that the distribution of services was one of 'unpatterned inequality'. Services were certainly unevenly distributed but the inequalities were not cumulative for particular areas or particular social groups. Inertia was one of the principal factors explaining parks, museums and fire protection services with newer peripheral areas receiving relatively fewer services. Finally, Jones and his associates conclude from their analysis of environmental services in Detroit that, as in other North American cities, there is no overall pattern of consistent discrimination against the poorer areas which would support any simple form of 'underclass hypothesis' (Jones *et al.*, 1980).

Examples of such studies could be repeated many times but the crucial question for studies of territorial justice is how to interpret such findings. Do they represent a striking refutation of the 'underclass hypothesis' or is there some methodological weakness which limits the range of their insights? The answer must be some combination of these explanations depending upon the type of the study.

Needs Resources and Outcomes

Certainly there are deficiencies in many of these studies in terms of the way in which they have measured key variables. For example the concept of inverse care embodies some notion of the criteria which warrant care — or in other words need, and this is also the central concept governing the allocation of resources within the idea of territorial justice (Pinch, 1979). The concept of need remains one of the most problematic issues in the social sciences and the full implications of methodological and philosophical issues surrounding the concept cannot be examined here. It is sufficient to note however, that the idea of need has often been only crudely examined in studies of service allocations within cities. Frequently the main interest has been upon the racial socio-economic or neighbourhood dimensions of variations without a detailed examination of whether these neighbourhoods need more police patrols, libraries, education facilities and the like. Some North American studies claim to perceive no discrimination against black areas in the sense that all areas obtain roughly equal amounts of service but this fact alone is of little importance without due consideration of underlying social conditions. In the medical field the derivation of health indicators has been largely dependent upon mortality statistics. These display a strong class bias – the lower your social class the greater your chances of dying. However, it is less clear how these mortality rates are related to patterns of disease, for people do not necessarily die from the diseases they suffer. The changing patterns of diseases brought about by modern sanitation and the advent of antibiotics is no longer reflected in mortality figures (Doyal, 1979). Recognising these difficulties there have been several attempts to collect morbidity figures – a task which Logan (1972) describes as 'movement from relatively firm but difficult ground to a treacherous bog'. Parallels may be drawn with needs indices in service studies.

Indices of housing need for example, almost inevitably rely upon census-based data concerning the presence of standard amenities and the extent of overcrowding. These measures appear to work reasonably well in isolating the tracts of pre-1914 terraced housing with their lack of amenities but, as Taylor (1979) points out, some of the worst 'slums' in Britain today exist in the pre-war local authority housing estates. Although these new housing units have the full range of standard amenities, many of them, and especially the high-rise blocks, are associated with a whole range of social and environmental problems (Pacione, 1982).

Faced with the enormous difficulties of measuring needs many assessments of inverse-care in the medical field have examined patterns of patient utilisation of health facilities (OPCS, 1973). These studies present a fairly consistent pattern in which the lower socio-economic groups visit their doctors more often. In the absence of rigorous indices of need however, these data are extremely difficult to interpret and this has led to some tortuous debate. Rein (1968) for example, accepts that mortality ratios may be extremely weak guides to health needs but notes that variations in consultation rates between social classes are much less than variations in death rates. There then follows the classic *reductio ad absurdum* in which he assumes that the consultation rates can be used as a needs index thus avoiding confounding the true incidence of class morbidity with the differential class response to illness. However, this argument fails to take account of all the complex factors which can affect the extent to which individuals consult their doctors. It is likely that better education makes the higher socio-economic groups more aware of their symptoms but lower socio-economic groups may have a greater need for medical certificates.

Measuring inverse care by general practitioner consultation rates leads to a situation in which morbidity is defined as a condition requiring medical treatment. This runs counter to the logic of the idea of territorial justice which attempts to make a clear analytic and conceptual distinction between the needs of areas and the resources which are allocated to meet these needs. This is important because a crucial distinction affecting the size of a problem is whether it is measured in terms of a problem identified throughout a population or whether it is measured in terms of the help given to alleviate the problem. As in the medical case there are examples of service studies which compound needs with studies of provision levels. Hatch and Sherrott (1973) for example attempted to derive indices of deprivation at a ward level for the London boroughs of Southwark and Newham. Despite their recognition that 'the level of provision may effect policy as much as need', they included amongst their need variables the number of persons admitted to accommodation for the homeless and the number of meals-on-wheels served.

Hart's original formulation claimed that the availability of *good* medical care tended to vary inversely with the need for it in the population. In effect, the quality of care (implied by the term good) has often been poorly measured in studies of service provision. (For further discussion of the problems see Davies, 1968; Pinch, 1979, 1980.) This deficiency is compounded by two important characteristics of service studies. First, most studies of territorial or spatial justice are based upon areas which are aggregates of individuals. The ecological fallacy means that such patterns of allocation between areas do not necessarily imply similar patterns amongst individuals within the areas, for one cannot be certain that those in most need actually receive the service. Second, and more important is the lack of information on *outcomes* — the effects of services upon the well-being of individuals. It is likely that such information on outcomes would go some way towards explaining the apparent paradox of non-cumulative inequality in service provision between areas but continuing social, economic and environmental deprivation. Rich (1982) observes that in the North American context studies have concentrated upon 'basic' public services such as police, fire protection, sewers and street lighting which confer roughly equal benefits to a wide section of the public. This ensures that public expenditure essentially maintains and reinforces the existing broader inequalities of wealth in society. Even where the services have redistributive objectives frequently they do not begin to cope with the magnitude of the problems posed by the poorest areas or their underlying causes. This is the case even when compensatory schemes are designed to overcome the difficulties of the worst areas. Thus the provision of public housing by itself does not overcome the poverty which prevents many families from adequately looking after their homes, hospitals by themselves cannot cope with many of the underlying causes of ill-health, while the provision of additional police resources may not affect the underlying causes of crime.

There are therefore two important requirements for future studies of territorial justice. The first is an expansion of the range of phenomena studied so that they accord more closely with Harvey's original notion of real income. For example, we need to look at such issues as the ways in which planning regulations enforce property rights and capital accumulation (Simmie, 1981), the processes affecting the location of noxious facilities, transportation policy and the activities of the police upon social control. Ultimately this will take studies away from the analysis of local services *per se* towards an understanding of the ways in which both public and private sector institutions affect welfare outcomes in particular localities. A second and related requirement is the need for more theory of the causes of inequality and a greater understanding of the links between services and local welfare outcomes.

Pluralism

Pluralism is now one of the oldest perspectives used to explain the distribution of services in cities, stemming from the late fifties and early sixties (Dahl, 1956; Polsby, 1963). In fact, it represents the modern embodiment of the liberal-democratic view of local government which can be traced back to the ideas of John Stuart Mill. He recognised that in modern industrial societies with a complex division of labour, a comprehensive system of participatory democracy is impossible and it is therefore necessary to delegate political power to representatives who are accountable to an educated and informed electorate at periodic elections (Goldsmith, 1980). Pluralists argue that within such a system of representative democracy politicians respond to the demands made by many different pressure groups. In essence this view is based upon an analogy with private markets as portrayed in neo-classical economics. Indeed, pluralism is but one variant of a set of ideas commonly known as public choice theory which attempt to apply economic concepts to the domain of political science (Archer, 1981; Reynolds, 1981). Local politics is thus envisaged as a political marketplace in which politicians respond to the preferences of the public. It is argued that the stronger the preferences the more likely it is that groups will take action and the greater the actions the more likely that politicians will respond. Pluralists argue that these responses do not reflect any overall class bias for the state is a neutral arbiter between competing interests. Broad equality of outcomes is guaranteed by the diversity of interest groups. Thus individuals will find themselves in many different and non-overlapping groups according to different issues. Furthermore, overriding all of these activities are the ever present elections through which politicians may be brought to account. Viewed superficially the pluralist idea of 'dispersed inequalities' would seem to fit with the observation of non-cumulative inequalities in service provision. It is also undeniable that in recent years local politics has involved a remarkable increase in pressure group activity (Kirk, 1980). Nevertheless, the pluralist view has been severely criticised in recent years (see Dunleavy, 1977; Lukes, 1974; Saunders, 1975, 1979).

From the point of view of service provision one of the most frequent criticisms is that the approach fails to take account of the widely differing political 'purchasing power' of various pressure groups. It would seem that frequently the more affluent and better educated sections of society are able to organise themselves more effectively to get their demands implemented. In contrast many deprived or disorganised groups with limited resources fail to get their objectives into the arena for political debate. For example, there is evidence of considerable dissatisfaction with the level of pre-school provision for the under fives in Britain (Bone, 1977), but mothers are a dispersed group with little political powers. The pluralist perspective portrays the local state as a passive receiver of demands from the public and ignores the role governments can play in formulating public opinion. Dearlove's (1973) research indicates the ways in which local councillors can resist demands from interest groups with whom they disagree. Furthermore the pluralist approach places an emphasis upon political activity.

Political inactivity is interpreted as satisfaction with the *status quo* or else low preference levels. This of course ignores the fact that many people may not participate because they feel their demands will not be recognised.

The fact that in many spheres all groups receive roughly equal amounts of service *despite* the observation of unequal levels of political participation finally highlights the overall weakness of the approach. Although local service provision may in some instances be affected by consumer demands, it would seem that often this is not the only or even the most important factor at work. For example, Jones and associates (1980) observed that in Detroit there was an absence of organisational activity concerned with service outcomes. There was certainly sporadic activity by community and neighbourhood groups but sustained pressure of the type necessary to explain service outcomes was apparently not the norm. Such evidence has led researchers to concentrate upon other determinants of service provision — notably the internal-decision-making mechanisms of local bureaucracies (via managerialism) and the external constraints under which local governments operate (via structuralism).

Managerialism

The clear differences between the nature of social organisation in Britain and North America have been responsible for very different intellectual traditions within urban geography. This began to manifest itself in the early 1970s with the emphasis upon the manner in which the housing market dictated residential patterns, in so far as the unique mixture of public and private forms of market organisation in Britain nudged several researchers towards an explicit analysis of the way in which the market was actually run: in short, the emphasis shifted from the study of *demand* which is as we have seen the keystone of neo-classical economics and public choice theory, to the study of *supply*.

This shift of emphasis could have taken several forms, but research in geography (and indeed throughout urban studies) has been heavily influenced by Pahl's reflections on what he termed 'urban managers' (1970), a contribution which has already been extensively reviewed (see in particular Saunders, 1980, pp. 166–80; Saunders, 1981, pp. 118–36; Pahl, 1975, Ch. 13 and Williams, 1982, pp. 95–105).

As we shall see, an emphasis upon gatekeepers — managerialism — is now seen as rather facile (it has also been termed both 'naïve' and 'elitist'), but it is important not to overlook the enormous gap (a veritable giant step) that existed between urban geography in, say, 1970 and 1975. Within a relatively brief period, a new generation of geographers turned their backs on the ecological legacy and its siblings: factorial ecological studies, morphological models and computer-based analyses derived from the Lowry model (Knox, 1982, pp. 112–39). Their interests focused relatively quickly upon the ways in which decisions were taken; particularly decisions which produced social injustices and spatial maldistribution, such as the relegation of 'problem families' to particular local authority housing estates noted for their bad design, or the direction of Building Society funds away from the inner areas and those in greatest housing need (Boddy, 1976; Gray, 1976; see also other papers in this volume). Similarly, in North America Levy and associates (1974) and Lineberry (1977) both argued that the decision rules adopted by local bureaucrats in public organisations were the single most important factor affecting the distribution of local public services — a view which has been supported by many others (Lipsky, 1974; Jones *et al.*, 1980). This research however proved to be intermediary in its impact; it provided not the basis for an understanding of decision-making so much as a glimpse into the complicated realities of social and political forces. The realisation of these forces almost immediately drew research down a path that promised to explain not simply the process of residential differentiation, but nearly everything else in the universe as well.

It is perhaps easiest to illustrate the problems associated with a managerialist perspective if we attempt to apply it to a particular case study. Our example focuses upon the role of the police in dealing with criminal offenders.

Crime and Punishment

Figure 9.1 shows the distribution of serious offences known to the police in England and Wales standardised by population, for 1980. It can be seen that there exists a basic rural-urban distinction in terms of crime rates, which is a consistent finding (Herbert, 1982). The highest rates of recorded crime are in London, the West Midlands (Birmingham), South Wales (Swansea, Cardiff), Merseyside (Liverpool), Greater Manchester (Manchester, Salford), West Yorkshire (Leeds, Bradford), Humberside (Hull), Cleveland (Middlesborough), Northumbria (Newcastle-Gateshead) and Nottinghamshire: all reflect centrality, except for the latter, which is anomalously high; (on the relationships between urban size and negative externalities, see Richardson, 1973).

Figure 9.1: Distribution of Serious Offences per 100,000 Population by Police District (1980)

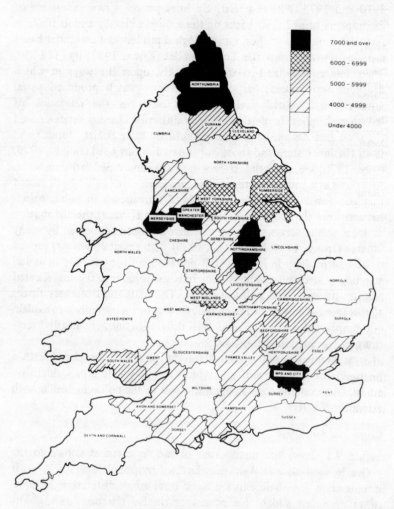

When we turn to the distribution of punishment, we would expect *either* that this would be in step with crime rates (that is high crime rates would be reflected in deterrent sentencing) *or* that procedures would be standardised. In fact, neither seems to prevail, as Table 9.1 shows. As the data indicate, in some police areas (including some with high crime rates), quite large proportions of those arrested never appear in court at all, and this proportion varies by a factor of about two.

Table 9.1: Males aged 10-17 Cautioned as Proportion of all Males 10-17 found Guilty or Cautioned

Police Area	Caution %	Police Area	Caution %	Police Area	Caution %
Avon	42	Humberside	31	Surrey	48
Bedfordshire	53	Kent	45	Sussex	48
Cambridgeshire	41	Lancashire	44	Thames	49
Cheshire	40	Leicestershire	43	Warwickshire	39
Cleveland	35	Lincolnshire	51	W. Mercia	56
Cumbria	39	Merseyside	42	W. Midlands	45
Derbyshire	40	Metropolitan	38	W. Yorkshire	39
Devon & Cornwall	65	Norfolk	55	Wiltshire	52
Dorset	53	Northamptonshire	44	Dyfed-Powys	63
Durham	40	Northumbria	49	Gwent	40
Essex	57	N. Yorkshire	40	N. Wales	46
Gloucestershire	48	Nottinghamshire	55	S. Wales	31
Greater Manchester	38	S. Yorkshire	41		
Hampshire	59	Staffordshire	50	England & Wales	
Hertfordshire	42	Suffolk	56	Mean	40

Source: Extracted from GB Criminal Statistics, 1981, Table 5.6.

(This question of variation can also be applied at later stages in the judicial process; Herbert, for example, clearly indicates that sentencing also varies from police force to police force in its severity: 1982, pp. 12-13.)

Why is it that some police forces appear to be more lenient than others? A managerial perspective would suggest that this will be a function of the way in which the individual police forces operate, although this does not determine exactly how such an assertion can be tested. As Williams notes 'managerialism is not a theory, nor even an agreed perspective. It is instead a framework for study' (1978, p. 236); in other words, all managerialism can tell us is that we should focus upon decision-makers — but who are they in this context?

One possibility in this instance is to concentrate upon the attitudes of individual police officers. Studies at the scale of police districts within cities show that the chances of a juvenile being charged or treated more leniently vary with the race of the suspect: 'black juveniles seem to be treated by the police more severely than their white counterparts' (Landau, 1981, p. 42). This does not 'prove' of course that variations in treatment reflect *individual* attitudes (a point which will be returned to below), although studies of police officers do reveal consistent tendencies. Colman and Gorman, for example, show that police recruits tend to hold extreme views on issues like immigration and mixed marriage; 'in my opinion most niggers especially Rasters should be wiped out of distinction' (1982, p. 7). These attitudes are ameliorated during basic training, when liberal ideas are presented, although ultimately a reversion to original attitudes occurs: 'socialisation into the police subculture seems therefore to foster hostile attitudes towards coloured immigration. The depth of prejudice towards black immigrants held by many experienced police officers can be gauged from the responses quoted above' (Colman and Gorman, 1982, p. 9; it is of interest to note that the latter is a Detective Chief Inspector).

Of course, these remarks underline the problem of examining individuals in isolation, and even the problem of knowing *which* individuals to examine. A consistent criticism of managerialism has questioned whether one should focus upon the 'top-dogs' or the 'middle-dogs'; in this instance, the police officers or the senior staff (Norman, 1975). It is hardly coincidence that Chief Constable Alderson of Devon and Cornwall has spoken consistently about the importance of liberal police attitudes, and that Devon and Cornwall has the *highest* rate of cautions in England and Wales, or that Chief Constable Anderton of Greater Manchester, who has called repeatedly for the reintroduction of capital punishment and the removal of local authority controls over police activity and funding, leads a police force with one of the *lowest* rates of cautioning (Table 9.1).

To extend this argument somewhat further, it is unsatisfactory to overlook other influences upon decision-making. An obvious one is the Police Federation, which has consistently campaigned in the political arena for the return of the death penalty and an increased emphasis upon law and order (Reiner, 1982, pp. 469–71). The existence of the Federation is consistent with other groups which function to create a specific professional ethos, and we might argue that without such an ethos, there is little point in studying a particular group of professionals; in other words, without the Institute of Housing, or the Society of Education officers, there would be no consistency of decision-making in fields like housing management or education, and consequently nothing to study but individual – almost random – patterns of behaviour (Kirby, 1979; 1982a).

It will not have been overlooked of course that the concept of managerialism is now being seriously stretched in order to focus upon organisations. However, as Williams points out, it is exactly in this way that 'primitive managerialist' studies can be utilised. As already noted, autocriticism quickly pointed to the limitations of a focus upon decision-makers, admitting that whilst the latter were responsible for allocating scarce resources, they were not responsible for the determination of scarcity in the first instance. If anything is to be salvaged from a perspective upon managers, it must involve placing them back in some broader social context:

> in an examination of building societies one would need to erect a model of the role of credit within contemporary capitalism ... although the actions of societies can be empirically observed steps must be taken to view the societies less as organisations in themselves, but rather as a process around which conflicts and tensions exist. (Williams, 1982, p. 103)

Thus in the context of our study of crime and punishment, we have to place the police force and its members back within the functioning of society at large. We can, as does Miliband, see them as part of the process by which the state 'manages violence' (1973, p. 48), although given the localised funding and organisation of the police, it makes more sense to follow O'Connor in regarding them as part of the *local* state's emphasis upon 'social expenses'; those activities like the courts, the schools and the police which collectively maintain social order.

Two advantages stem directly from this wider interpretation. First, it allows us to explain the curious relationship between the state and the Police Federation; the latter is emphasising social order, whilst the former must also take into account the need to legitimise its actions — in order words, different priorities exist. Similarly, an emphasis upon the state reminds us of the 'role' of racism within capitalist society, as part of the means by which a particular underclass is created. Such an interpretation underlines the fact that racist police attitudes exist within a judicial and legal framework that are maintained by the state despite liberal political pressures and ethnic opposition.

This example has aimed to illustrate some of the problems that arise in attempting to understand spatial and social phenomena from a managerial perspective. As we can see, the problem quickly 'blows up', to take into account wider and wider issues. In fact, because of this, many workers in urban research have tended to move away from a study of human actions altogether, in order to focus directly upon the deeper structures within society.

Structuralism

Saunders observes that:

> today we are confronted with three different traditions in the study
> of urban politics which among them have given use to four competing
> theories: pluralism, which stresses the representativeness of state
> institutions; instrumentalism, which stresses the domination of the
> state by particular elite or class interests; managerialism, which
> stresses the political autonomy of those in key positions within
> the state; and what may, perhaps somewhat misleadingly, be termed
> structuralism, which stresses the necessary structural relationship
> between the state and dominant economic class interests in a capitalist
> society. (Saunders, 1982, p. 29)

Little will be said here for the moment concerning instrumentalism,
which has problems accounting for the relationships that exist between
elites and the state, in so far as we are left to 'explain' just what the
state is. This problem does not arise with a structuralist interpretation,
which emphasises the close functional connection between the state
and economic activity: in short, the two have evolved together under
capitalism.

Given the emphasis of this chapter, it is not possible to outline the
structural proposition in any detail. Of importance here is the realisation
that contradictions do exist between the state and capital, one of which
relates to the need for the former to reproduce the labour required by
the latter: in short, the state provides what capital will not. Thus
housing, education, health care and recreation are all financed and
allocated by the state, in accordance with capital's needs, user's
political demands and pressures, and the state's abilities to pay;
unsurprisingly, the outcome of these interrelationships is an overall
variation in provision, and the existence of shortage in some sectors and
some areas (Saunders, 1981, pp. 188-9). It should be noted that one
does not have to accept a Neo-Marxist interpretation to appreciate
that the State intervenes in many complex ways in modern industrial
societies to maintain order, legitimise inequalities and support productive
infrastructure. Indeed, it can be argued that there are strong similarities
between certain structuralist theories and functionalist 'system-type'
explanations. Both face similar problems in accounting for social
change without reference to the activities of agencies and agents. The
crucial issue is whether one perceives the state to act inevitably in the

long term interests of capital (as in structuralist view) or else in the interest of some idea of the 'common good' (however defined).

Unfortunately the structuralist perspective poses particular problems when attempts are made to operationalise it: any theoretical position which depends upon countervailing forces must necessarily be elastic in its 'prediction' of particular results and outcomes, as opposite extremes will reflect simply the success or failure of one or other protagonist. This then takes us into circular argument, where one empirical observation can be taken to represent the success of monopoly capital in determining state behaviour, whilst an opposite result can be taken as an indication that political opposition to the state has achieved some short-term advantage; like some epistemological game of football, any result — win, draw, or lose — can be accommodated into the expectations. Let us attempt to illustrate this problem with a particular case.

Library Provision in London. Mention has already been made of the state's role in the reproduction of the labour force. Such an interpretation of public provision is useful in so far as it permits a relatively direct explanation of the gross variations that occur between different spatial units. Thus, we would expect that demands for social investment and/or consumption goods (like education) will be resisted with greater vigour by political opponents and the state itself in areas where reproduction is of lesser importance: for example, in regions of traditionally low job skills, or inner urban areas, with high unemployment rates.

In some sectors, this argument can be empirically illustrated: in the fields of health, and education, broad patterns of investment commensurate with regional economic disparities can be identified (Kirby, 1982b; Chs. 2, 3). In other cases, the relationship is far more confused. Table 9.2 for instance outlines the regional variation in spending in England and Wales on libraries, a relatively important component of the educational process. In this case, expenditure appears to be highest in the regions with the most serious problems of economic decline (with the exception of London); such a reverse of our initial expectation would be difficult to account for without very detailed investigation (unless, as already noted, we introduce some notion of political gains in certain regions resulting in higher expenditure).

Problems arise even with studies at a far finer spatial scale, that is at the level of the individual decision-making units; Figure 9.2 for instance shows the expenditure upon libraries (only) in the London Boroughs, for 1979–80. It is clear that a distinction exists between

Table 9.2: Regional Variations in Expenditure on Libraries,[a] Museums and Art Galleries, England and Wales, 1980–1

Region	Expenditure per Head (£)
South-East	6.24
North-West	5.82
Yorkshire-Humberside	5.50
Northern	5.45
West Midlands	4.84
East Midlands	4.66
Wales	4.48
South West	4.40
East Anglia	4.08

Note: a. Including Museums and Art Galleries.
Source: CIPFA statistics.

levels of expenditure within the Inner London Boroughs and those on the periphery, which is again not consistent with accepted views of the functional relationship between service distributions and the economic system:

> state action not only facilitates class reproduction by maintaining social control, but also more directly contributes to that reproduction. Thus we should expect to find, for example, that the poorest schools are concentrated in low income and working class neighbourhoods while better schools dominate middle and upper income areas . . . that many services which would contribute to the economic mobility of the lower classes . . . are not provided or not provided in adequate quantity or quality . . . (Rich in Kirby and Pinch, 1982, p. 83)

Nor is there an explicable pattern of expenditure *cutting* in this sector. We can explain the tendency towards fiscal crisis (see Dear, 1981) and predict that its impact will be greatest where capital has the ability to manipulate the local state by opposing certain levels of public expenditure.

> at the most general level, it is capitalists who control city budgets, not local politicians, voters or muncipal employee unions. In pursuit of the economic interest of their individual firms, capitalists make choices which set limits to the size of the local public sector and dictate the mix of services it will contain. They need not conspire or intend to do this, or even be aware of the consequences of their actions for this to be true. (Rich in Kirby and Pinch, 1982, p. 78)

Figure 9.2: (a) Expenditure per head of Population in London on Library Books (1979-80); (b) Change in Expenditure on Library Books in London, Allowing for Inflation

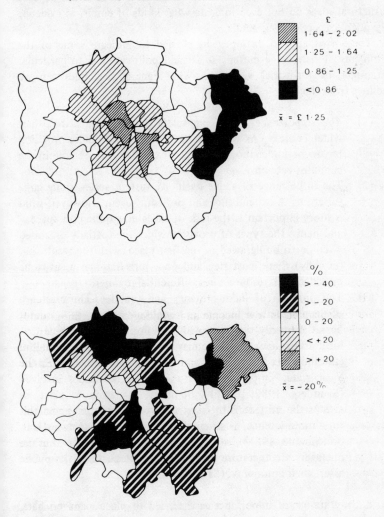

The extent to which such an interpretation is consistent with Figure 9.2 is again difficult to decide without detailed investigation; the expected patterns (perhaps of high status Boroughs being reluctant to withdraw provision in this field) could not be substantiated by this confused pattern.

This ultimately must be the limitation of what we have here termed a structural analysis – not that the argument cannot be empirically verified (for such a thing is inherently impossible), but rather because structural ideas do not even indicate what kinds of empirical evidence we might expect or seek out.

These two applied examples have tried to indicate some of the problems of harnessing current social and political ideas to particular examples of territorial variations. What general principles can we adduce from them?

(1)   The importance of placing spatial issues back within their wider context. As we have seen, the role of public provision has to be understood in the context of the state's role in the capitalist economy.

(2)   The importance of space itself. At various scales, quite large variations in expenditure and provision occur. However, what is more important is the task of explaining quite how space – and hence the types of provision which are spatially-organised – come to be jigsawed in the first place, with the result that (crudely) there exist rich and poor jurisdictional areas to be provided with resources in a differential manner.

(3)   The problem of linking theory and practice. The existence of deep structural tendencies – unlike earthquakes – cannot be satisfactorily demonstrated. It is implausible to argue that local affairs have no role to play; for example, instrumentalism (the link between business and the polity) is strong within the local state, simply due to the scale of the issues involved (Saunders, 1980, p. 211). Similarly, managers *do* exist: as always the mistake is to assert their total impotence, or their total omnipotence, as an emphasis upon corporatism tends to do (Simmie, 1981): 'theories that deny the relevance of urban managers are therefore every bit as inadequate as those that assert their autonomy' (Saunders, 1981, p. 128).

As already stated, of importance is the need to place managers back into our analysis in situations where they have a clearly identifiable role: perhaps in relation to dealing with central-local conflicts which become manifest as expenditure cuts in particular services or in particular localities.

Conclusions

This chapter has attempted to impose some order upon what is at times a seemingly chaotic mass of studies. This has inevitably led to over-simplication and distortion but it is possible to derive some clear lessons for future studies of territorial justice.

While geographers have in recent years been accused of making a fetish of space, it is undeniable that there are considerable persistent geographical variations in well-being at numerous scales. Indeed, it is the widespread recognition of these inequalities which has paradoxically led many other social scientists to enter the field of urban studies. Description and evaluation of these inequalities raises enormous problems but there is much which requires urgent analysis, especially at a time when fiscal policy in much of the western world is increasing inequalities. Two tasks are of particular importance. First there is the need to examine individual services in greater detail than hitherto, so that better indices of needs, service quality and outcomes can be derived. Second, there is a need to expand the range of phenomena studied so that we can obtain a better understanding of the distribution of real income in the city.

The most important task for the future, however, is to provide an adequate explanation for spatial variations in service provision. In this respect a major obstacle would seem to be the enormous multitude of factors and agents at work — local politicians, pressure groups, bureaucrats at various levels, trade unions, professional organisations, historical inertia, community structure and external constraints imposed by central government. Pahl (1979) argues that lists of 'factors' are no substitute for a comprehensive theory which can be used to understand the relationships between these numerous elements. However, as demonstrated in this chapter, the various pluralist, managerialist and structuralist interpretations emphasise the importance of particular features to the exclusion of others. Saunders (1980) argues that the pluralist and structuralist views are essentially ideological in character since their methodological presuppositions prevent any empirical repudiation of their main features. However, one does not have to accept the pluralist or structuralist viewpoints (with all their attendant ramifications) to appreciate that pressure groups and broader 'structural' constraints can affect local outcomes in various ways. After the explosion of theorising in the late seventies, there would now seem to be a growing awareness in urban studies that the influence of pressure groups, local managers and the central state is ultimately a matter for empirical validation rather than resolution at the theoretical level. This is not to argue for a return to the rightly discredited 'abstracted empiricism' of the sixties or to impute the belief that 'bits of theories' can be borrowed at random and welded into some *ad hoc* structure which can account for all circumstances. One crude, but possibly useful, starting point would be to recognise the importance of scale upon the processes at work. Thus if one is concerned with broad questions of resource allocation between sectors, one needs to take account of the broad social, economic and political environment within which decisions are made at the national level. If one is concerned with questions of spatial distribution between jurisdictions then these structural constraints are likely to form an important backcloth within which the activities of numerous local agents are likely to determine outcomes; as one moves closer to the local level then it seems likely that factors such as local pressure group activity may come into prominence. However, no simple distinction can be drawn between scale and process for individuals, groups and institutions at all levels work within the same macro-level constraints. Urban geographers in the eighties would therefore do well to take heed of the sentiments so well expressed by Lukes (1977).

Social life can only properly be understood as a dialectic of power and structure, a web of possibilities for agents, whose nature is both active and structured, to make choices and pursue strategies within given limits, which in consequence expand and contract over time. Any standpoint or methodology which reduces that dialectic to a one-sided consideration of agents without (external or internal) limits, or structures without agents, or which does not address the problems of their inter-relations, will be unsatisfactory.

References

Alt, J.E. (1973) 'Some Social and Political Correlates of County Borough Expenditures', *British Journal of Political Science*, *1*, 49-62
Alt, J.E. (1977) 'Politics and Expenditure Models', *Policy and Politics*, *5*, 83-92
Archer, J.C. (1981) 'Public Choice Paradigms in Political Geography' in Burnett, A.D. and Taylor, P.J. (eds.) *Political Studies from Spatial Perspectives*, John Wiley and Sons, Chichester
Bassett, K. and Short, J.R. (1980) 'Patterns of Building Society and Local Authority Lending in the 1970s', *Environment and Planning A*, *12(3)*, 279-301
Bennett, R.J. (1980) *The Geography of Public Finance*, Methuen, London

Boaden, N.T. (1971) *Urban Policy Making*, Cambridge University Press, Cambridge

Boddy, M. (1976) 'The Structure of Mortgage Finance: Building Societies and the British Social Formation', *Transactions of the Institute of British Geographers*, *NS(1)*, 58–71

Bone, M. (1977) *Pre-school Children and the Need for Day Care*, Office of Population Censuses and Surveys, London

Bradley, J.E., Kirby, A.M. and Taylor, P.J. (1978) Distance Decay and Dental Decay, *Regional Studies*, *12*, 529–50

Brazer, H.E. (1959) *City Expenditures in the United States*, Occasional Paper 66, National Bureau of Economic Research

Burnett, A.D. (1981) 'The Distribution of Local Political Outputs and Outcomes in British and North American Cities: A Review and Research Agenda', in Burnett, A.D. and Taylor, P.J. (eds.) *Political Studies from Spatial Perspectives*, John Wiley and Sons, Chichester, 201–35

Colm, G. (1936) 'Public Expenditures and Economic Structures in the United States', *Social Research*, *3*, 57–77

Colman, A. and Gorman, L.P. (1982) 'Conservation, Dogmatism and Authoritarianism in British Police Officers', *Sociology*, *16(1)*, 1–11

Dahl, R.A. (1956) *A Preface to Democratic Theory*, University of Chicago Press, Chicago

Davies, B.P. (1968) *Social Needs and Resources in Local Services*, Michael Joseph, London

Dawson, R.E. and Robinson, V.A. (1963) 'Inter-party Competition, Economic Variables and Welfare Policies in the American States', *Journal of Politics*, *25*, 265–89

Dear, M.J. (1977) 'Psychiatric Patients and the Inner City', *Annals of the Association of American Geographers*, *67(4)*, 588–94

Dear, M. (1981) 'A Theory of the Local State', in Burnett, A. and Taylor, P.J. (eds.) *Political Studies from Spatial Perspectives*, John Wiley and Sons, Chichester, 183–200

Dearlove, J. (1973) *The Politics of Policy in Local Government*, Cambridge University Press, Cambridge

Doyal, L. (1979) 'A Matter of Life and Death: Medicine, Health and Statistics' in Irvine, J., Miles, I. and Evans, J. (eds.) *Demystifying Social Statistics*, Pluto Press, London, 237–54

Duncan, S.S. (1974) 'Cosmetic Planning or Social Engineering? Improvement Grants and Improvement Areas in Huddersfield', *Area*, *9(4)*, 259–71

Dunleavy, P.J. (1977) 'Protest and Quiescence in Urban Politics: A Critique of some Pluralist and Structuralist Myths', *International Journal of Urban and Regional Research*, *1*, 193–218

Dye, T.R. (1966) *Politics, Economies and the Public Policy Outcomes in the States*, Rand McNally, Chicago

Fenton, J.H. and Chamberlayne, D.W. (1969) 'The Literature dealing with the Relationship between Political Processes, Socio-economic Conditions and Public Policies in American States', *Polity*, *1*, 388–404

Fisher, G.W. (1961) 'Determinants of State and Local Government Expenditures: A Preliminary Analysis', *National Tax Journal*, *17*, 64–73

Freestone, R. (1977) 'Provision of Child Care Facilities in Sydney', *Australian Geographer*, *13*, 318–25

Gray, F. (1976) 'Selection and Allocation of Council Housing', *Transactions of the Institute of British Geographers*, *NS10(1)*, 34–46

Great Britain (1981) *Criminal Statistics England and Wales 1980*, HMSO, London

Goldsmith, M. (1980) *Politics, Planning and the City*, Hutchinson, London

Harvey, D.W. (1971) 'Spatial Process, Spatial Form and the Redistribution of
    Real Income in an Urban System', in Chisholm, M., Frey, A.E. and Haggett, P.
    (eds.) *Regional Forecasting*, Butterworth, London, 270–300
Harvey, D.W. (1973) *Social Justice and the City*, Arnold, London
Harvey, D.W. (1977) 'Government Policies, Financial Institutions and
    Neighbourhood Change in United States Cities', in Harloe, M. (ed.) *Captive
    Cities*, Wiley and Sons, Chichester
Hatch, S. and Sherrott, R. (1973) 'Positive Discrimination and the Distribution
    of Deprivations', *Policy and Politics*, pp. 223–40
Hodgart, R.L. (1978) 'Optimising Access to Public Services: A Review of
    Problems, Models and Methods of Locating Central Facilities', *Progress in
    Human Geography*, *2(1)*, 17–48
Herbert, D.T. (1982) *The Geography of Urban Crime*, Longmans, London
Hofferbert, R.I. (1966) 'The Relation between Public Policy and some Structural
    and Environmental Variables in the American States', *American Political
    Science Review*, *10*, 73–82
Jones, B.D., Greenberg, J.R. and Drew, J. (1980) *Service Delivery in the City*,
    Longman, New York
Kirby, A.M. (1978) *Education, Health and Housing. An Empirical Investigation
    of Resource Accessibility*, Saxon House, Farnborough
Kirby, A.M. (1979) 'Managerialism and Local Authority Housing, A Review',
    Public Administration Bulletin, *30*, 47–60
Kirby, A.M. (1982a) 'Education, Institutions and the Local State' in Flowerdew,
    R. (ed.) *Institutions and Geographical Patterns*, Croom Helm, London
Kirby, A.M. (1982b) *The Politics of Location: An Introduction*, Methuen,
    London
Kirby, A.M. and Pinch, S. (1982) *Urban Politics and Public Provision*, GP80,
    University of Reading
Kirk, G. (1980) *Urban Planning in a Capitalist Society*, Croom Helm, London
Knox, P.L. (1978) 'The Intra-urban Ecology of Primary Medical Care: Patterns
    of Accessibility and their Policy Implications', *Environment and Planning*,
    A10, 415–35
Knox, P.L. (1982) *Urban Social Geography*, Longmans, London
Landau, S.F. (1981) 'Juveniles and the Police', *British Journal of Criminology*,
    21(1), 27–46
Lasswell, H.D. (1958) *Politics, Who gets What When and How?*, World Publishing
    Co., New York
Le Grand, J. and Winter, D. (1977) 'Toward an Economic Model of Local
    Government Behaviour', *Policy and Politics*, *5*, 23–9
Levy, F.S., Meltsner, A.J. and Wildavsky, A. (1974) *Urban Outcomes: Schools,
    Streets and Libraries*, University of California Press, Berkeley
Lineberry, R.L. (1977a) *Equality and Urban Policy*, Sage Publications, Beverley
    Hills
Lineberry, R.L. (1977b) 'Equality, public policy and public services: the
    Underclass Hypothesis and the Limits to Equality', *Policy and Politics*, *4*,
    67–84
Lipsky, M. (1974) 'Toward a Theory of Street Level Bureaucracy' in Hawley,
    W.D. and Lipsky, M. (eds.) *Theoretical Perspectives in Urban Politics*, Prentice
    Hall, New Jersey
Logan, R.F. (1972) 'Dynamics of Medical Care', Liverpool Studies into the Use of
    Hospital Resources, London School of Hygience and Tropical Medicine
Lukes, S. (1974) *Power, A Radical View*, Macmillan, London
Lukes, S. (1977) *Essays in Social Theory*, Macmillan, London
Margolis, J. (1968) 'The Demand for Urban Public Services' in Perloff, H.S. and

Wingo, L. (eds.) *Issues in Urban Economics*, John Hopkins Press, Boston, 535–56

Margolis, R. (1977) 'Ratfields, Neighbourhood Sanitation and Rat Complaints in Newark, New Jersey', *Geographical Review*, *67(2)*, 221–32

Miliband, R. (1973) *The State in Capitalist Society*, Quartet, London

Mladenka, K.R. and Hill, K.Q. (1977) 'The Distribution of Benefits in an Urban Environment: Parks and Libraries in Houston', *Urban Affairs Quarterly*, *13(1)*, 73–94

Mladenka, K.R. and Hill, K.Q. (1978) 'The Distribution of Urban Policy Services', *Journal of Politics*, *40*, 112–33

Morrill, R., Earickson, R.J. and Reis, P. (1970) 'Factors Influencing the Distance Travelled to Hospitals', *Economic Geography*, *46*, 161–71

Musgrove, R.A. (1959) *Theory of Public Finance*, McGraw-Hill, New York

Newton, K. and Sharpe, L.J. (1977) 'Local Outputs Research: some Reflections and Proposals', *Policy and Politics*, *5*, 61–82

Nicholson, R.V. and Topham, N. (1971) 'The Determinants of Investment in Housing by Local Authorities: An Econometric Approach', *Journal of the Royal Statistical Society*, *Series A 134*, 273–320

Nivola, P.S. (1979) 'Distributing a Municipal Service: A Case Study of Housing Inspection', *Journal of Politics*, *40*, 59–81

Norman, P. (1975) Managerialism: 'A Review of Recent Work', *Conference Paper 14*, Centre for Environment Studies, London, 62–86

O'Connor, J. (1973) *The Fiscal Crisis of the State*, St Martins Press, New York

Office of Population Censuses and Surveys (1973) *The General Household Survey*, HMSO, London

Pacione, M. (1982) 'Evaluating the Quality of the Residential Environment in a Deprived Council Estate', *Geoforum*, *13(1)*, 45–55

Pahl, R.E. (1970) *Whose City?*, Longman, London

Pahl, R.E. (1975) *Whose City?*, 2nd edn., Penguin Books, Harmondsworth

Pahl, R.E. (1977) 'Managers, Technical Experts and the State' in Harloe, M. (ed.) *Captive Cities*, John Wiley and Sons, Chichester

Pahl, R.E. (1979) 'Socio-political Factors in Resource Allocation' in Herbert, D.T. and Smith, D. (eds.) *Social Problems and the City*, Oxford University Press, London, 33–47

Pinch, S.P. (1978) 'Patterns of Local Authority Housing Allocation in Greater London between 1966 and 1973: An Inter-Borough Analysis', *Transactions of the Institute of British Geographers (NS) 3*, *(1)*, 35–54

Pinch, S.P. (1979) 'Territorial Justice in the City: A Case Study of the Social Services for the Elderly in Greater London' in Herbert, D.T. and Smith, D.M. (eds.) *Social Problems and the City*, Oxford University Press, London, 201–23

Pinch, S.P. (1980) 'Local Authority Provision for the Elderly: An Overview and Case Study of London' in Herbert, D.T. and Johnston, R.J. (eds.) *Geography and the Urban Environment vol. 3*, John Wiley, London, 295–343

Polsby, N. (1963) *Community Power and Political Theory*, Yale University Press, New Haven

Rein, M. (1969) 'Social Class and the Health Service', *New Society*, *14*, 20 November

Reiner, R. (1982) 'Bobbies take the Lobby Beat', *New Society*, *59*, 469–71

Reynolds, D.R. (1981) 'The Geography of Social Choice' in Burnett, A.D. and Taylor, P.J. (eds.) *Political Studies from Spatial Perspectives*, John Wiley and Sons, Chichester, 91–109

Rich, R. (1979) 'Neglected Issues in the Study of Urban Service Distributions', *Urban Studies*, *10*, 143–56

Rich, R. (1982) 'Urban Development and the Political Economy of Public

Production of Services' in Kirby A.M. and Pinch, S.P. (eds.) *Urban Politics and Public Provision*, GP80 University of Reading

Richardson, H.W. (1973) *The Economics of Urban Size*, Saxon House, Farnborough

Sacks, S. and Harris, R. (1964) 'The Determinants of State and Local Government Expenditures and Inter-government Flows of Funds', *National Tax Journal*, *17*, 75–85

Samuelson, P.A. (1954) 'The Pure Theory of Public Expenditure', *Review of Economics and Statistics*, *36*, 387–89

Saunders, P. (1975) 'They Make the Rules', *Policy and Politics*, *4*, 31–58

Saunders, P. (1979) *Urban Politics: A Sociological Interpretation*, Hutchinson, London (reprinted 1980, Penguin, Harmondsworth)

Saunders, P. (1981) *Social Theory and the Urban Question*, Hutchinson, London

Saunders, P. (1982) Community Power, Urban Managerialism and the Local State, in Harloe, M. (ed.) *New Perspectives on Urban Change and Conflict*, Heinemann, London, 27–49

Seley, J.E. (1979) 'A Comparison of Technical and Ethnographic Approaches to the Study of Fire Services', *Economic Geography*, *59*, 36–51

Sharkansky, I. and Hofferbert, R.I. (1969) 'Dimensions of State Politics, Economics and Public Policy', *American Political Science Review*, *63*, 867–79

Sharrard, T.D. (1968) (ed.) *Social Welfare and Urban Problems*, Columbia University Press, New York

Simmie, J.M. and Hale, D.J. (1978) 'The Distributional Effects of Home Ownership and Control of Land use in Oxford', *Urban Studies*, *15*, 9-21

Simmie, J.M. (1981) *Power, Property and Corporatism*, Macmillan, London

Smith, D.M. (1977) *Human Geography: A Welfare Approach*, Edward Arnold, London

Taylor, P.J. (1979) 'Difficult-to-let, Difficult-to-live-in, and sometimes Difficult-to-get-out-of': An Essay on the Provision of Council Housing with Special Reference to Killingworth, *Environment and Planning*, *A11*, 1305–20

Thompson, L.R. (1965) *A Preface to Urban Economics*, John Hopkins University Press, Boston, p. 118

Tudor-Hart, J. (1971) 'The Inverse Care Law', *The Lancet*, February, 405–11

Webster, B.A. (1981) 'The Distributional Effects of Local Government Services' in Leach, S. and Stewart, J.D. (eds.) *Approaches in Public Policy*, George Allen and Unwin, London

Williams, P.W. (1978) Urban Managerialism: A Concept of Relevance? *Area*, *10*, 236–40

Williams, P.W. (1982) 'Restructuring Urban Managerialism: Towards a Political Economy of Urban Allocation', *Environment and Planning*, *A14(1)*, 95–195

10 POLLUTION

D.M. Elsom

A pollutant is a substance or effect which adversely alters the environment by changing the growth rate of species, interferes with the food chain, is toxic, or interferes with health, comfort, amenities, or property values of people (Holister and Porteous, 1976). Pollution may be regarded as a form of resource misuse arising because individuals or groups transfer the costs of dealing with their waste products to others. It is most obvious in urban areas where man's socio-economic activities have concentrated his waste-forming activities.

The Nature of Pollution

Wastes discharged into the environment may take a variety of forms so discussion of pollution is often separated according to the environmental medium through which the pollutants diffuse: air, water or land. Most national pollution control policies have been developed along this compartmentalised approach being concerned with remedying the severe deterioration of each medium resulting from the waste-products of industrialisation and urbanisation. Frequently the strategy for pollution control has been to establish environmental standards and to adopt tactics so that these standards are met. Such an approach has suffered problems. For example, efforts to meet water quality standards may lead to efficient removal of pollutants from industrial effluents or domestic sewage, so as to discharge potable water, but the pollutants have not been eliminated. The waste-products have been merely altered in form and moved from one place to another. In the case of water treatment plants the process has produced a sludge. If it is incinerated, the risk of air pollution exists. If the sludge is spread on the land the risk of contaminating vegetation, crops and the food chain with toxic materials exists (Collin, 1978). How does one make the choice? Such considerations highlight that the pollution problem must be tackled ultimately in a holistic and preventative manner. This has led to a second type of pollution control strategy as illustrated by the 'Environmental Impact Assessment' legislation introduced first by the

United States but subsequently adopted, or in the case of European nations, being considered for adoption (Chapman, 1981). Such an approach attempts to prevent the emergence of a serious pollution or environmental problem from a planned development. However, as with compartmentalised pollution control policies its effectiveness is being reduced because the current international economic recession is persuading many governments to relax or refrain from enforcing environmental policies which are perceived to retard economic growth and increase inflation and unemployment.

To illustrate the development of various approaches to the pollution problem and the perceived changes in the nature of that problem, the following sections examine the air pollution problem for selected national control strategies. Air pollution was selected rather than land or water pollution as studies in urban communities during the past two decades have highlighted that air pollution was perceived as the prime pollution problem (Dworkin and Pijawka, 1981; Hewitt and Burton, 1971; Saarinen and Cooke, 1970; Van Ardsol *et al.*, 1964).

Air Pollution

Atmospheric pollution originates from gaseous and particulate wastes emitted into the air as a consequence of the burning of fuels used in the processing and production of materials, the powering of vehicles, the production of electricity, and the space-heating of building. Common gaseous pollutants include oxides of sulphur, nitrogen, and carbon, hydrocarbons and ozone while particulate pollutants, solid matter which may remain suspended in the atmosphere for some time, include smoke, dust and grit. The composition of such particulates varies from unburnt carbon to complex substances including lead and radioactive compounds. Given adequate dilution and dispersion in the atmosphere urban and industrial emissions of common pollutants may not be regarded as giving rise to significant adverse effects. The recognition of the harmful effects or the undesirable nature of the pollutants arises when meteorological conditions, often aggravated by local topography, fail to disperse or dilute the emissions leading to high concentrations of pollutants at ground level. Resulting air pollution episodes, during which concentrations reach socially unacceptable, economically costly and unhealthy, even dangerous, levels may last a few hours or several days (Brodine, 1971; Elsom, 1978a, 1979). Such episodes are typically of two types: smokey-sulphurous smogs and photochemical smogs.

Smokey-sulphurous smogs result from the burning of fossil fuels (coal, oil, and gas) for the generation of electric power, for industrial processes, and for space-heating. During cold anticyclonic conditions in the winter emissions of pollutants escalate as space-heating of homes, offices and factories becomes essential but the atmosphere, characterised by only light winds and a restricted mixing layer, has limited capability of dispersing and diluting the emissions. The presence of vast quantities of hygroscopic nuclei leads to the formation of fog as temperatures fall and relative humidity increases. Although the pollution concentrations within the smog are largely a function of the local emissions the pollutants may be spread throughout the urban area. In urban areas the city centre is frequently a few degrees warmer than the outskirts (the heat island effect) and this may lead to the generation of a thermal breeze circulation, analogous to a sea-breeze effect. This mesocirculation takes the form of the warmed central air rising and moving out to the suburbs while cooler air from the outskirts is drawn into the city centre. The result is that the pollutants are spread throughout the urban area (Elsom, 1978b; Padmanabhamurty and Hirt, 1974). Emergency measures such as requiring industry to change to sulphur-free fuels or even to shut down may be adopted but the smog will remain for as long as the anticyclone persists, which may be several days.

Whereas it is the stationary sources of pollution which are responsible for the smokey-sulphurous smogs, it is the automobile which is held responsible for photochemical smogs. This smog was first recognised in Los Angeles in the 1940s where it was identified as being produced by the action of sunlight on hydrocarbons and oxides of nitrogen present in vehicle exhausts. The photochemical smog consists of ozone, aldehydes, and peroxyacetylnitrate (PAN), collectively referred to as 'oxidants' which cause eye irritation, severe swelling of the mucus membrane thus restricting inhalation and causing coughing, damage to plants, and reduced visibility. Although recognised in the 1940s in the United States, it was not until the 1970s that photochemical pollution was perceived as a problem in other major cities of the world. In cities such as Tokyo, Sydney and London, photochemical pollution emerged as a problem with startling suddenness. As motor vehicle exhaust emissions of hydrocarbons and oxides of nitrogen increased so did the likelihood that the Los Angeles type of smog would be experienced. During the hot sunny summers of 1975 and 1976 European cities reported peak hourly concentrations of ozone exceeding 20 pphm, the level at which a first-stage alert is initiated in Los Angeles. This first-stage alert, called on average between one to six times every year since the 1960s in Los Angeles, is a health warning but if levels exceed 35 pphm (second-stage alert) active measures are taken to control emissions (Lin, 1981). During a second-stage alert in Los Angeles on 14 July 1978 all industries which produced hydrocarbons and oxides of nitrogen were required to reduce their output by 20 per cent, oil-burning power stations were forced to switch to natural gas or imported electricity, households were requested to switch off electrical appliances, including the air-conditioning despite the 35°C temperature, residents were advised to avoid strenuous outdoor exercise, and commuters were asked to share car transport to work. Fortunately, the third-stage alert or emergency situation, set at ozone levels of 50 pphm, during which industry could be closed down and traffic brought to a halt has yet to be called (Whittow, 1980).

A variety of emergency measures may be taken by city authorities to reduce the adverse effects of smogs but the solution to the pollution problem lies in long-term pollution control strategies. Today, nearly all pollution control strategies throughout the world are based on either air quality management or emission standards, or a combination of both (Thornes, 1979). Economic strategies in which taxes, charges, or fees are applied to air pollution emissions have also been suggested (Rosencranz, 1981). However, pollution charges depend upon the ability to measure accurately the quantity and nature of the emissions from each individual emission source, and so, apart from Czechoslovakia where some success has been achieved using emission taxes, economic strategies largely remain theoretical. Similarly, cost/benefit approaches to pollution control are unlikely to be adopted at present because our knowledge and expertise is too inadequate to allow us to quantify all the many terms involved in cost/benefit analyses of pollution and pollution control (Berry and Horton, 1974). For example, the costs to such aspects as amenities is presently so speculative as to be of little real value (Chandler, 1976).

Air quality management is defined by de Nevers *et al.* (1977) as the regulation of the amount, location, and time of pollutant emissions to achieve some clearly defined set of ambient air quality standards or goals. Advocates of the alternative emission standard strategy argue that since the air quality management approach requires calculating the emission levels needed to achieve air quality standards, why not shortcut the development of ambient air standards, air monitoring, diffusion modelling, etc. and simply require all emitters to go directly to a 'good practice' emission standard (de Nevers *et al.*, 1977). Emission standards are being used more and more in the United States whereas in Britain they represented the complete control strategy until the recent European Communities pollution control directive which required Britain to adopt health protection air quality standards. The development of approaches to pollution control in the United Kingdom, European Communities, United States of America, and Socialist countries such as China, are examined in the following sections.

The British and European Community Approaches to Pollution Control

The development of air pollution control in Britain was strongly influenced by the occurrence of a London smog or 'pea-souper' on 5–8 December 1952 (Ashby, 1975). This smokey-sulphurous smog character-ised by dangerously high pollution concentrations at a time of below-freezing temperatures led to a dramatic rise in the number of deaths, especially among the elderly, due to bronchitis, influenza, pneumonia, tuberculosis, and other respiratory illnesses. An estimated 4,000 excess deaths were attributed to the smog although to put this in perspective, an estimated 12,000 Londoners were destined to die in January and February 1953 from an influenza epidemic and those who died during the smog would have been those most likely to have died a month later from the influenza epidemic (Perry, 1981). Nevertheless, the emotive statement that 4,000 deaths were attributable to the smog provided the trigger for the media and the public to press the government for air pollution control legislation. Low-level emissions of smoke from domestic open-fires were viewed as the major cause of smogs. Some local authorities, notably Manchester and Coventry, had already recognised this and had established smoke control areas within their urban areas in which only smokeless fuels were allowed to be burned. An increasing number of authorities had intended to introduce private bills to Parliament to create similar smoke control areas but as a result of the 1952 smog disaster were asked to await the outcome of the Government committee investigation into the nation's air pollution problems. Following the committee's report in 1954 the Government were slow to act but were pressed into action by the proposed sub-mission of a private members Clean Air (Anti-Smog) Bill in 1955. This bill was withdrawn on the promise of a Government bill being intro-duced. The Clean Air Act was passed in 1956 and granted local authorities the power to demarcate 'smoke control areas' in which only authorised manufactured smokeless fuels (for example, types of coke, gas and oil) could be burned. Because these fuels could not be burned in existing domestic grates, the Act provided government grants to be paid to householders to help defray the cost of purchase and installation of new heating applicances (Scarrow, 1972). The Act made smoke control programmes a matter of local option rather than statutory duty. Given that the local authorities were required to contribute 30 per cent to the cost of the necessary household heating conversions (central government contributed 40 per cent and the property owner 30 per cent) this resulted in the larger, wealthier authorities such as London, Yorkshire and parts of northwest England adopting a progressive programme while other authorities were slow to respond. Cost was not always the most important influence on the pursuit of establishing smoke control areas. In coal mining areas there was a strong lobby against establishing smoke control areas. Miners received a concessionary coal allowance and designation of smoke control areas would restrict its use. Moreover, Wall (1973) showed that South Yorkshire miners believed pollution restrictions would reduce the demand for coal so effecting redundancies and the policy might also discourage the introduction of new industries into the area.

Given the regional variation in the designation of smoke control areas central government took the power to require local government action through the Clean Air Act of 1968. Over 300 polluted or 'black' areas were deemed to be priority cases for pollution abatement (Scarrow, 1972). However, the government has not always encouraged pollution control. In 1970–71 many smoke control areas were temporarily suspended because of shortages of smokeless fuels and this led to a worsening of the air pollution problem in some urban areas (Elsom, 1979). More recently economic cut-backs by central government have led to conversion grants being withdrawn and between December 1976 and June 1977 a moratorium was imposed on new smoke control area orders (Department of Environment, 1979). Economic recession during the 1970s has led to many local authorities repeatedly deferring the target dates for completion of their urban smoke control area programme. By 1978, eight million premises were covered by smoke control orders (Department of Environment, 1979).

The Clean Air Acts of 1956 and 1968 were not only concerned with domestic smoke emissions but with industrial smoke emissions too. Industry was pressed to employ the 'best practicable means' to reduce their emissions and was required to adopt minimum heights for chimneys. Overall, the approach to industrial polluters was one of persuading them to adopt a code of 'good emission conduct': very few prosecutions of industry have taken place.

Government, the media and the public have frequently praised the Clean Air Act of 1956 together with its amendment and strengthening in 1968. The First Report of the Royal Commission on Environmental Pollution (1971) concluded 'since the first Clean Air Act became law in 1956 there has been a steady reduction in the emission of smoke and sulphur dioxide into the air over Britain', and it warned that 'the downward trends in smoke and sulphur dioxide pollution are encouraging, but will continue *only if* there is no relaxation in applying the provisions of the Clean Air Acts ...' Auliciems and Burton (1973) expressed reservations concerning such strong claims and pointed out that the 1956 Clean Air Act in particular was merely 'swimming along with the social, economic and technological tide'. Indeed, smoke had steadily decreased since the 1920s and probably since the late nineteenth century (Brimblecombe, 1977, 1978, 1982). The traditional British image of a 'cosy coal fire' was changing as affluence followed the post war depression. A demand for higher heating standards meant that the cleaner and more efficient systems which used solid smokeless fuels, oil and gas (North Sea sulphur-free natural gas became available after 1967) were welcomed. Advertising media stimulated consumer preference for the clean efficient central heating systems. Slum clearance programmes replaced dense terraced housing characterised by multiple low-level emission sources with multi-storey dwellings having a central efficient heating system. Bernstein (1975) and Scarrow (1972) attempted an approximate assessment of the contribution of the Clean Air Acts to the improvement of London air quality. Both concluded that the Acts helped to finance smoke control, through grants for heating appliance conversions, for only 15 to 30 per cent of London householders. For other less prosperous areas of Britain, especially the north, percentages were higher but all reveal that social, economic and technological factors were significant in reducing pollution levels.

Whatever the relative importance of the causes, smoke pollution in urban areas decreased dramatically since the 1950s. By the 1970s, urban populations were enjoying the benefits of less suspended particulate matter in the air in terms of a healthier, sunnier and less foggy atmosphere. In the 1940s and 1950s the number of hours of winter sunshine in city centres such as London and Manchester was 50 per cent that of the surrounding rural area but by the mid-1970s this difference had almost disappeared (Department of Environment, 1979; Tout, 1979). This improvement has taken place partly due to the increase in atmospheric transparency but mostly through the reduction in the frequency and duration of fogs as particulate matter encourages the formation of fogs even well below 100 per cent relative humidity (Brazell, 1970; Harris and Smith, 1982; Unsworth *et al.*, 1979).

Whereas smoke concentrations in urban areas have decreased by 80 per cent during the past 20 years, decreases in other pollutants have been less marked or have even increased. This is primarily because the Acts tackled visible air pollution, especially as expressed in smogs, and the invisible gaseous pollutants, while ranking as harmful, received less attention. Pollutants such as sulphur dioxide are less easy to detect by the general public and are largely overlooked. Wall (1974a, 1974b, 1976) found in Sheffield that between 1949 and 1971 complaints mentioning only particulates varied between 75 and 96 per cent compared to 2 to 13 per cent mentioning gaseous pollutants at all. Even so, levels of sulphur dioxide have decreased, if at a lesser rate than smoke, during the past two decades which may confirm the importance of non-legislative influences on emission reductions. The reduction of smoke pollution has also helped to reduce sulphur dioxide concentrations by lessening the frequency of fogs which encourage the build up of other pollutants.

In some areas such as London and Sheffield, whereas smoke control is almost complete, sulphur dioxide pollution remains a problem. For example, in December 1975 meteorological conditions of a very stable temperature profile, low wind speeds and temperatures below freezing point produced a pollution episode with peak smoke concentrations of 811 $\mu g/m^3$ and sulphur dioxide concentrations of 1,238 $\mu g/m^3$ (Apling *et al.*, 1977). Fuel oil was believed to be the major source of the high sulphur dioxide concentrations and this had already led the Greater London Council in 1972 to require all new oil-fired installations to use fuel-oil with a maximum of one per cent sulphur content. Under Section 76 of the Control of Pollution Act 1974 (the sections 75–84 covering air pollution control being brought into force in January 1976) other local authorities are permitted to do this.

The 1975 London air pollution episode revealed not only that sulphur dioxide pollution remains a problem but that despite 93 per cent of Greater London being covered by smoke control orders smoke pollution can still reach undesirable levels. This confirms that motor vehicles, rather than domestic emissions arising from coal consumption, are the single most important source of smoke pollution (Ball and Hume, 1977). If smoke concentrations were to be further reduced in London it is the vehicular source of particulates, on average about 75 per cent of 'dark smoke', and during pollution episodes as much as 90 per cent, which must be controlled (McGinty, 1977). The increased number of vehicles has also given rise to significant increases in urban concentrations of lead, carbon monoxide, hydrocarbons and oxides of nitrogen. It is the vehicular exhaust emissions of hydrocarbons and oxides of nitrogen that have been held responsible for the recent increase in photochemical pollution in London and the rest of the United Kingdom. During 1975 and 1976 peak hourly ozone levels in London reached 15 and 21 pphm respectively (Thornes, 1977). Although London may have experienced the worst photochemical pollution in the United Kingdom, on occasions ozone levels were simultaneously high throughout the country (Derwent *et al.*, 1976). This confirmed the belief that ozone (and other pollutants such as sulphates and nitrates) was being imported from continental Europe (Ball and Bernard, 1978; Barnes and Lee, 1978; Cox *et al.*, 1975).

By the 1970s it became clear that the air pollution problem, once perceived primarily as urban smogs arising from domestic smoke emissions, had become a complex problem involving diverse pollutants from stationary and mobile emission sources. Further, the scale of the problem had changed. Long-distance transport of pollutants was now considered a significant contributor to the problem. The strategy adopted by the United Kingdom and other European countries to reduce local ground level pollution concentrations from industrial emission sources was to increase the height of industrial stacks. This policy reduced local concentrations but enabled pollutants to travel long distances. Oxides of sulphur and nitrogen released in this way transform to sulphates and nitrates before reaching the ground and these forms of the pollutants can travel even further before they are finally scavenged from the atmosphere by precipitation. This process increases the acidity of the precipitation and 'acid rain or snowfall' can lead to a serious environmental problem in some countries. For example, Southern Sweden, an environment lacking the acid-neutralising or 'buffering' capacity of limestone-based soils and waters, has suffered serious acidification of over 20,000 lakes threatening the survival of its freshwater fauna and flora. Scandinavian countries, being subjected to a high frequency of winds carrying pollution from Britain and northwest continental Europe, are net importers of pollution and acid precipitation and have pressed for international action to reduce this trans-boundary pollution problem (Barnes, 1979). However, although the United Nations Organisation have charged nations not to cause damage to environments outside their borders and to develop international laws to safeguard against trans-boundary pollution no remedial action has been taken. Few countries show a willingness to meet the costs of the action necessary to remove the pollutants at source or to accept interference in their national pollution control policies (Rosencranz, 1980). The European Community is the exception to this situation as the ten member states are compelled to follow EEC pollution policy. However, the basis for a common EEC pollution strategy lies less with a concern for trans-boundary pollution as a concern for fair trade between member states. Pollution control measures can account for 20 to 30 per cent of the cost of some industrial processes. Varying pollution control policies between countries can create barriers to fair trade within the EEC so for this reason unification of pollution control strategies has been pursued since 1973 (Johnson, 1979).

The EEC are convinced there is sufficient medical evidence to propose health protection air quality standards for various pollutants. In July 1980, the EEC issued a directive for the maximum concentrations of smoke and sulphur dioxide permitted in urban areas (Table 10.1). In recognition of the synergistic effect of these two pollutants on health the health protection standard for sulphur dioxide is dependent on the level of smoke present: the more smoke that is present on a particular time scale, the less the amount of sulphur dioxide allowable (Thornes, 1979). Other pollutants were to be tackled later. This strategy represented a major change in the British approach to pollution control and Britain expressed its opposition to the imposition of air quality standards (Royal Commission on Environmental Pollution, 1976). British strategy had relied heavily on 'reasonableness', on the willingness of people to employ the 'best practicable means' to reduce pollution to some desired environmental quality level rather than apply coercive penalties (Burton *et al.*, 1974). In contrast to the British approach most other members of the EEC have serious doubts whether a system based on goodwill and voluntary compliance are adequate, especially where dangerous and toxic substances are involved. They believe more in standards defined on the basis of best technical means and applied through mandatory instruments (Levitt, 1980).

Member states have until 1983 to comply with the standards, with the tactics for implementing the standards being left to the individual states. The EEC directive emphasised that the implementation of the standards in urban areas must not lead to a deterioration of air quality in the 'clean' regions; as far as possible, compliance with the standards must be achieved by reducing emissions and not by wider dispersal of pollutants in the environment (Johnson, 1979). It is not clear as to the

Table 10.1: European Community Directive Concerning Health Protection Ambient Air Quality Standards for Sulphur Dioxide and Suspended Particulates in Urban Atmospheres

Sulphur dioxide

Reference period	Maximum concentrations	Associated concentrations of suspended particulates
Year	Median of daily means 80 μg/m³	Annual median of daily means > 40 μg/m³
Year	Median of daily means 120 μg/m³	Annual median of daily means < 40 μg/m³
Winter (October to March)	Median of daily means 130 μg/m³	Winter median of daily means > 60 μg/m³
Winter (October to March)	Median of daily means 180 μg/m³	Winter median of daily means < 60 μg/m³
24 hours	Arithmetic mean 250 μg/m³	Arithmetic mean of concentration over 24 hours > 100 μg/m³
24 hours	Arithmetic mean 350 μg/m³	Arithmetic mean of concentration over 24 hours < 100 μg/m³

Suspended particulates

Reference period	Maximum concentrations
Year	Median of daily means 80 μg/m³
Winter (October to March)	Median of daily means 130 μg/m³
24 hours	Arithmetic mean 250 μg/m³

Source: Johnson (1979).

legal rights the EEC has to enforce such a directive and what penalties are to be given for non-compliance. The EEC will be dependent on member states supplying information from sampling and monitoring sites established throughout each country. Since 1961, the United Kingdom has had an extensive smoke and sulphur dioxide monitoring network, the National Survey of Air Pollution, which has consisted of approximately 1,200 sites. As from April 1982 this network was drastically rationalised to form the United Kingdom Smoke and Sulphur Dioxide Monitoring Network of 150 sites (Elsom, 1982; Handscombe and Elsom, 1982). The reason given for this rationalisation was that smoke and sulphur dioxide concentrations had fallen so low that costly monitoring was no longer necessary. An alternative reason is that such a dense network increases the number of individual sites which may indicate failure of one or more of the EEC standards. As the Greater London Council (1979) stressed, 'it would be unfair if we and other European cities with extensive monitoring networks had to undertake greater expenditure to reduce pollution levels than less carefully monitored cities'. In spite of this comment British cities are currently required to maintain some 200 sites, in addition to the rationalised national network of 150 sites, in areas where there is a possibility that the EEC standards may be exceeded in the next few years.

The adoption of the EEC health protection air quality standards for smoke and sulphur dioxide in urban atmospheres opens the door for further air quality standards. EEC directives are being prepared for lead, oxides of nitrogen, carbon monoxide, and hydrocarbons. Such a strategy for air pollution control, a major departure for the traditional British approach to pollution control, follows the strategy adopted by the United States of America over a decade ago.

Pollution Control in the United States of America

Whereas in most European countries smoke and sulphur dioxide have been accorded priority for air pollution control during the past two or three decades, a wide range of air pollutants was tackled very early on in the United States pollution control programme. While most European countries considered domestic and industrial emitters of pollutants as the prime target for pollution control, in the United States the automobile was recognised to require similar attention as a source of air pollution. Concern for vehicle exhaust emissions of pollutants originated in Los Angeles in the 1940s when the dirty yellow smog which blanketed the city for most of the summer daylight hours was shown to be produced by the action of sunlight on hydrocarbons and oxides of nitrogen. Having highlighted the diverse nature of the air pollution problem in urban atmospheres, the United States approach to pollution control was to establish National Ambient Air Quality Standards (NAAQS) for seven pollutants — particulate matter, sulphur dioxide, carbon monoxide, hydrocarbons, oxides of nitrogen, photochemical oxidants (ozone), and lead (Table 10.2).

NAAQS were established by the 1970 Clean Air Act which followed the less comprehensive legislation of the Air Pollution Act of 1955, the Clean Air Act of 1963, the Motor Vehicle Air Pollution Control Act of 1965, and the Air Quality Act of 1967. Primary and secondary NAAQS were set by the Environmental Protection Agency for the entire nation rather than leaving it to individual states. Each state was required to develop plans to achieve the primary standards, designed to protect

Table 10.2: National Ambient Air Quality Standards in the United States of America

Pollutant	Period of Measurement[a]	Primary Standard		Secondary Standard	
		$\mu g/m^3$	pphm	$\mu g/m^3$	pphm
1. Carbon monoxide	8 hours	10,000	900	Same	Same
	1 hour	40,000	3,500	Same	Same
2. Hydrocarbons (non-methane)	3 hours	160	24	Same	Same
3. Nitrogen oxides	year	100	5	Same	Same
4. Ozone	1 hour	240	12	Same	Same
5. Sulphur oxides	year	80	3	None	None
	24 hours	365	14	None	None
	3 hours	None	None	1,300	50
6. Total suspended particulates	year	75	—	60	—
	24 hours	260	—	150	—
7. Lead	3 months	1.5	—	None	None

Note: a. Standards for periods of 24 hours or less may not be exceeded more than once per year, except ozone may use three year statistical average to determine if exceeded.
Source: Purdom (1980).

public health, within a 'reasonable time' (Wetstone, 1980). Following some relaxation of the deadline in the early 1970s, the 1977 Clean Air Act Amendments require that all states attain the primary NAAQS by 1982 with an extension for ozone and carbon monoxide until 1987. However, it is already evident that many states will not achieve attainment by 1982. For example, the South Coast Air Basin in California would need to reduce gasoline consumption by over 30 per cent to bring that area into compliance for oxides of nitrogen (National Environmental Development Association, 1981).

More stringent secondary NAAQS which aim to safeguard aesthetics, vegetation and materials will follow attainment of the primary NAAQS. Those regions where air quality was as good or better than the primary NAAQS to begin with were required to move directly to secondary NAAQS. Outlying areas where the air quality was even better than secondary NAAQS should not allow 'significant deterioration' of this quality. This Prevention of Significant Deterioration (PSD) policy imposes a limit on the overall increase in pollution concentration allowable in an area.

As well as air quality standards, emission standards have been adopted for motor vehicles. Post-1975 vehicles are required to reduce emissions of carbon monoxide, hydrocarbons, and oxides of nitrogen to 90 per cent of 1970 or 1971 vehicle levels (Brodine, 1972). Similarly the removal of the anti-knock gasoline additive of lead has been given priority with new cars being introduced in 1975 designed to use lead-free petrol which is required to be made available at all petrol stations.

The collective effect of the adoption of air quality and emission standards is to make the United States air pollution control programme 'one of the most aggressive in the world' (Wetstone, 1980). Even so, some assessments of the effectiveness of the Clean Air Acts point to much of the improvement in the air quality since the 1960s being attributed to a limping economy and the continuing substitution of clean fuels (oil and natural gas) for coal. If this is so, as the economy expands and national energy policy forces a return to coal, air pollution could get markedly worse (Marshall, 1981). A contributing factor to air quality improvement since the 1960s has been the reduction in the frequency of adverse meteorological conditions which produce winter smogs (it will be interesting to assess the effects of a succession of cold winters beginning in the late 1970s on the trend of space-heating emissions of pollutants). To suggest that air pollution in urban areas may worsen in the near future is especially worrying given that many urban areas have not yet attained primary NAAQS and some urban

Figure 10.1: Non Attainment Areas for the National Ambient Air Quality Standards for Total Suspended Particulates, August 1977

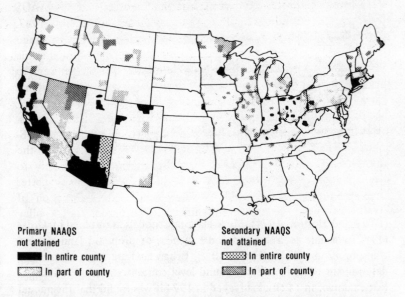

Primary NAAQS
not attained
■ In entire county
▨ In part of county

Secondary NAAQS
not attained
▧ In entire county
▨ In part of county

areas contain locations where pollution greatly exceeds the NAAQS as shown in Figure 10.1 (Berry, 1977; Murchett, 1981; Purdom, 1980).

Under the Clean Air Act of 1970, the basic tool for attaining and maintaining air quality standards is the State Implementation Plan (SIP). Each state must present monitoring and modelling data indicating that its control programme will bring about the attainment of the NAAQS. If the state fails to develop a satisfactory plan or execute it, the Environmental Protection Agency has the authority to intervene directly although with its budget cut by 12 per cent in 1983 it may lack the necessary resources to do so (Heneson, 1982). The models employed to develop the control programme vary from the simple 'proportional' or 'rollback' model that assumes, for example, a region with sulphur dioxide levels twice the NAAQS will attain the standard if total sulphur dioxide emissions in the region are halved, to more complex diffusion or dispersion models. Dispersion models predict ambient concentrations of pollutants from emission inventories of pollution sources, meteorological conditions, and topographical considerations. Such models not only allow various emission control strategies to be examined for existing pollution sources but can assess the likely impact of new pollution sources

Figure 10.2. Air Quality Simulation System

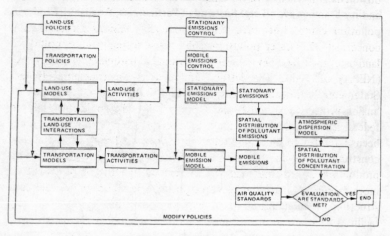

(Berry and Horton, 1974). More elaborate models as outlined by Gross (1982) are able to assess the consequences of proposed land-use and transportation policies (Figure 10.2). Urban land-use types are a major determinant of the resulting ground level concentrations of pollutants. For example, in 14 cities surveyed in 1974 it was found that the annual mean concentration of particulate matter was 50 to 70 $\mu g/m^3$ for residential monitoring locations, 60 to 110 $\mu g/m^3$ for commercial monitoring sites, and 80 to 150 $\mu g/m^3$ for industrial monitoring locations (Murchett, 1981). Modelling land-use and transportation policies represent the first stage of the Air Quality Simulation System (Gross, 1982). The output from this stage is passed to the emission stage of the model where, based on emission characteristics, the transportation and land-use activities are converted to pollution emission rates. The output from this stage represents the geographical distribution and intensity of emissions over the region. Based on this information, meteorological and atmospheric dispersion models are employed to predict the spatial distribution of pollution concentrations over the metropolitan region.

The spatial distribution of large industrial emission sources of pollutants within urban areas is an important consideration in a pollution control programme and it is pertinent to note that pollution control measures are influencing industrial locational and development decisions. For example, with new industrial plants having to comply with stringent technology-based New Source Performance Standards (NSPS), some companies prefer to continue operating old plant or to buy an existing facility rather than build a new, more tightly controlled, and more expensive to operate unit (Wetstone, 1980). It is not only air pollution control measures which influence decisions but legislation concerned with water pollution, solid waste disposal, noise pollution, land-use, and the comprehensive National Environmental Policy Act (NEPA) of 1969. The latter requires that Environmental Impact Statements be researched, and submitted, before the construction of 'major projects'. Although the NEPA is directly applicable only to federal agencies or projects requiring federal licences, in practice it has been extended to all major schemes of industrial expansion and new construction. Given that manufacturers frequently prefer to expand production at existing facilities, either through the employment of more labour or machine, or through multiple shifts, or via the construction of an addition to the plant, rather than investing in new facilities in a new location the new environmental regulations may well reinforce this preference for in-site expansion for several reasons (Stafford, 1977):

(1) Larger operations can more readily absorb environmental control costs due to attainment of economies of scale in waste disposal. Starkie (1976) referred to a survey of industries in the United States revealing that most committed in excess of 5 per cent of total capital expenditure for pollution abatement in 1973. Some industries committed far more such as the paper industry (43 per cent), non-ferrous metals (23 per cent), iron and steel (14 per cent), and chemicals (12 per cent). In 1978, capital expenditure for air pollution control alone was estimated by Hart (1982) to have cost industry $7.5 billion; operation and maintenance costs in the same year rose to over $9 billion. These figures compare with the $8 billion a year saved in wages and productivity which would have been lost due to sickness caused by air pollution and possibly $51 billion for the annual benefits (reduced medical treatment; less damage to buildings, crops and forests) to the nation as a whole of a less polluted society (Costle, 1979).

(2) It is probably easier to get permission to expand production in an area where one is already operating especially given that the additional discharge of wastes would represent a smaller percentage increase in pollution levels at the existing location than at a new site. However, restrictions on expansion in already heavily polluted or very densely populated areas have increased. For example, 186 out of the 247 United States Air Quality Control Regions (AQCRs) have not attained air quality standards for at least one of the pollutants and development

may be barred under current regulations. To overcome this federal restriction, most states that are not in compliance with a federal air quality standard have adopted the 'offset' policy (McAfee, 1982). Any company that wants to build or expand a plant that will emit pollution must first arrange for a greater reduction of emissions from existing local sources. This can be achieved by a company reducing emissions from its own plant in the area, by paying another company to reduce its emissions, or by purchasing an old plant in the area and simply closing it down. Increasingly industrialists are finding offsets difficult to obtain and are urging Congress to allow some relaxation of this policy (National Environmental Development Association, 1981). Environmentalists argue for its retention and believe the offset policy recognises that air is a scarce resource and uses market forces to accommodate growth without increasing total pollution (National Clean Air Coalition, 1981). Recognising that industrialists should have some say in what pollution measures are adopted by their companies the Environmental Protection Agency introduced the 'bubble' concept in 1979. Regulations previously required a company to bring each item of its plant — foundries, ovens, boilers, or whatever — into line with strict control levels. The bubble concept treats the entire plant, and sometimes a series of plants, as a single polluter and requires a cutback in the aggregate amount of fouled air. This leaves the company to decide whether it is cheaper and more efficient to control pollution from one valve or smokestack instead of another (Smith, 1981a).

(3) Having to comply with environmental regulations, monitoring pollution levels in alternative proposed locations for at least a year, and employing elaborate computer models to demonstrate that pollution from its proposed plant will stay within the allowable pollution increment, searching for offsets, etc., all increase the time and cost of locating a new facility as well as introduce a degree of uncertainty. Delays may arise because concerned citizens opposed to the planned facility in a specific location may challenge it through the judicial process. Whereas previously a site for a new plant could have been selected in three to six months it may take two or more years (Stafford, 1977). By that time industrialists argue it may be too late to get into the market place with the product. Further, it may cost $250,000 to $300,000 (McAfee, 1982). Various models have been developed to assess the locational response of industry to pollution control policies although the complex nature of the policies have yet to be modelled realistically (Coppock, 1974).

Harris (1981) argues that the frequent claims for the negative impacts of environmental policies on the business community conceal several positive impacts. While pollution control costs have put over 20,000 people out of work due to factory closures, pollution control is a major growth industry and has created close to one million jobs between 1971 and 1977 (it has also created a lobby of industrialists who favour further pollution control!). Those factories which went out of business because they lacked the necessary capital to improve pollution control measures were frequently the older inefficient plants. Many operations were transferred from these plants to more modern and efficient factories in the southern states, where energy costs are lower, labour unions are less organised, and industrial expansion and transportation are not confined by the intense land utilisation characteristic of the north-east. Overall, Harris (1981) believes the impacts of environmental policies may have been negative at the local scale and in the short-term but positive at the national scale and in the long-term.

In general, the perceived negative nature of some environmental policies has led to increasing pressure by industrialists for a relaxation of certain policies. Some success has been achieved with for example, the automobile industry winning major concessions on emission requirements in the 1977 Clean Air Act Amendments. Further, in January 1979, following sustained political pressure from the American Petroleum Institute, the ozone NAAQS was relaxed from 8 pphm to 12 pphm, saving the petroleum industry billions of dollars otherwise needed to comply with the former standard. Even this relaxation of the standard, shifting 10 to 20 large cities from the dirty to the clean list, still leaves cities such as New York, Washington DC, Houston, Denver and Los Angeles in need of major ozone control programmes (Marshall, 1978, 1979). In Los Angeles the ozone problem appears intractable and the 1987 deadline for compliance is likely to be missed (Smith, 1981b). Whether industry will be penalised for non-attainment is unclear. Industrialists will continue the pressure for further relaxation of environmental regulations and for the extension or removal of deadlines for compliance with NAAQS. When the economy is weak and industrialists complain that pollution control requirements have driven up prices, reduced labour forces, and weakened their competitiveness compared to foreign competitors it is not surprising that governments relax environmental regulations. Unlike the 1961–71 period, unemployment and inflation currently outrank pollution as social and political issues. With the Clean Air Act of 1970 (and the Clean Water Act of 1972) up for reauthorisation in 1982–3 it may be that the 1980s will be remembered as a decade of setbacks in the desire for continually improving environmental quality.

Approaches to Pollution Control in Socialist Countries

The pollution problems which characterise western societies may be perceived by some to arise because economic production is based principally on private ownership and profit-making. This may suggest that the pollution problems in Socialist countries are less serious and that industrial and urban growth is more compatible with notions of conservation, environmental protection, and improving the quality of life (Sandbach, 1980). However, it appears that this is not the case whether one considers the Soviet Union, the eastern bloc of European countries, or China.

In the Soviet Union, although by the 1960s individual republics had adopted a variation of the 'Law of the Preservation of Nature', there is no national environmental law covering pollution control (Komarov, 1980a). Large-scale industries such as iron and steel, chemicals, mining and power, who pollute the environment more than anyone, are accountable only to the central government (in which pollution control agencies have a low standing in the ministerial hierarchy) so the environmental laws in the republic in which the industry is located are not binding on the industrialists. The desire for economic growth has overridden environmental considerations and pollution is a serious problem (Komarov, 1980a, 1980b).

The pollution problem is of similar severity in Poland even though national pollution standards have been adopted. In the Katowice region, the nation's industrial heartland and source of most of its coal and steel, such vast quantities of particulate matter are discharged into the atmosphere that every urban area in the region experiences more than the 'permitted' annual fallout of 250 tonnes per square kilometre, with many cities receiving four times this level, and with Cracow receiving nine times the national standard (Timberlake, 1981). Gaseous pollutants are converted to acid precipitation which corrodes buildings, damages vegetation, and encourages respiratory illnesses. Pollution may be a violation of Polish law but the bureaucratic complexities and desire for economic growth lead to little action in terms of factory closure, prosecution, or pressure to incorporate pollution control equipment. Indeed, the national emphasis on industrialisation is such that efforts are directed more towards adapting the environment to industry, such as planting deciduous trees to replace sensitive Scotch pines, rather than altering industrial processes so as to control sulphur dioxide emissions (Kormondy, 1980). The Solidarity Trade Union Movement may have lifted the censorship on information on pollution for a brief period, and even achieved the closure of a major polluting aluminium smelter (Rich, 1981), but its demise and the economic predicament of the nation mean that tackling pollution is far down the list of priorities, and little change is likely in the near future. Not all Socialist countries rank the environmental quality issue at a similar level as practised in the Soviet Union and Poland. Kormondy (1980), Kromm (1973) and Kromm *et al.* (1973) suggest that environmental awareness and the desire for pollution control is much higher in Hungary and Yugoslavia as indeed it is in China.

In contrast to the Soviet Union and European Socialist countries, China has adopted a development policy based upon decentralisation and relying to a great extent on small-scale or intermediate technology. Scarcity of resources for industry has encouraged recycling, or multi-purpose use as the Chinese prefer to call it (Sandbach, 1980). A campaign launched in the 1970s to urge people to make the maximum use of waste products (materials, water, gas, and heat) has not only benefited the economy but reduced the discharge of harmful waste products into the environment (Huanxuan, 1980). With a planned economic policy resting upon a strategy aimed at improving the quality of life rather than being governed by market forces and the requirement to maximise profits, pollution should be less of a problem than in capitalist or even some other socialist countries. Indeed, the official Chinese line taken a few years ago was 'to lecture the despicable West about its environmental mess and about China's enviable policies of thoughtful protection and care' (Smil, 1980a). In contrast, the current official line is to admit the sorry plight of the country's environment. Simple small-scale technology suffers from inefficient energy conversion which together with the widespread use of low-quality solid fuels (bituminous coal, lignite, peat, and oil shale) emitted from low-level emission sources lead to a serious air quality problem (Smil, 1980b). Even though 80 per cent of China's 838 million people (in 1975) live in the countryside, which reduces the urban sprawl that could have resulted from such a large population, large urban and industrial areas exist and the pollution problem in China's northern cities is severe. The Maoist industrial policy developed in the 1960s involved not only the use of small-scale technology but also modern large-scale technology for certain key products such as chemicals, oil and power. Whereas emissions of particulates can be partially controlled using simple equipment, controls to reduce gaseous emissions such as sulphur dioxide from large-scale industries are inadequate. The current plans for fast and large-scale industrialisation in the coming decades will triple the mass of sulphur dioxide entering the atmosphere with a consequent escalation of the acid precipitation problem which during the 1970s was clearly identified as the main cause of acidity in fresh water and soil over extensive areas.

Current concern for pollution problems resulted in the strengthening of the Environmental Protection Office in 1979 so that it could adopt a more substantive policymaking role (Boxer, 1981). Numerous industrial enterprises throughout the country were ordered to take 'proper measures' to control pollution by 1982 at the latest or be closed down. The timing and extent of this action are clear indications of the seriousness of the pollution problems and it reveals that China faces pollution at least as serious as in any large modernising nation (Smil, 1980a, 1980b).

Conclusion

Pollution is a consequence of the past and present imbalance in the importance given to economic growth and technological advance on the one hand and safeguarding environmental quality on the other. Economic growth and technological advance have brought considerable benefits to mankind but the benefits have often been bought at the cost of a deterioration of our environment especially noticeable in the quality of urban environments where man has concentrated his waste-forming activities. Pollution is a problem which all societies currently suffer or will be confronted with eventually. Societies which favour decentralisation and small-scale technology may experience less pollution than those with extensive urbanisation and large-scale industries but differences between societies are being reduced as international and global pollution increases. Many national pollution control programmes lead to increased trans-boundary pollution as this effectively transfers the cost of the pollution to other countries. As a result some countries have become net-importers of international pollution as a consequence of their geographical position.

Increasing recognition, especially beginning in 1969, that economic growth is not an end in itself but only a means to promote human welfare, provided the impetus for the establishment or strengthening of national pollution control programmes and for international conferences to consider the problem of international and global pollution. The pollution control programmes initially developed by countries were, and in many cases are still, piecemeal and curative. Pollution of the environmental media of water and air have received more attention than land pollution so it is not surprising that problems, such as the legacy of toxic waste disposal, have emerged subsequently as major pollution problems. Attempting to control only selected pollutants, while disregarding or overlooking others, has led to societies having to face a diverse range of major pollution problems concerned with, for example, pesticides, oil spills, noise, thermal waste, and radioactive waste. This situation emphasises that a comprehensive or holistic and preventative approach to pollution control is needed.

Some of the drive to increase the importance of safeguarding environmental quality against the consequences of economic growth has been lost since the emergent years of the environmental movement in 1969-71, especially in the face of current economic depression, inflation and unemployment, but it has to be regained. Deterioration of the environment as a result of discharging wastes today may be a permanent deterioration which cannot be remedied or it will involve the next generation in a costly curative programme. The implication of such a statement has been exemplified recently by the discovery of dangerous toxic waste dumps in New York State which will require 20 years of remedial work costing forty times what it would have originally cost to have 'safely' disposed of the wastes in the 1940s. However, there was no remedy for the illnesses, miscarriages, birth defects and despair experienced by communities near the dumps. A curative approach to pollution control is no substitute for prevention. Prevention ultimately implies, in this case, reshaping the basis of our chemical technology on the principle of environmental compatibility. A similar argument applies to other pollution problems such as when the issues of radioactive waste and the nuclear power industry are substituted for toxic wastes and the chemical industry. Whether a successful preventative pollution control policy will be adopted by developed and developing countries ultimately depends upon political and ideological considerations and the willingness of societies to accept the social and economic adjustments necessary.

References

Apling, A.J., Keddie, A.W.C., Weatherley, M-L.P.M. and Williams, M.L. (1977) 'The High Pollution Episode in London, December 1975', *Report LR 263 (AP)*, Warren Spring Laboratory, Stevenage

Ashby, Lord E. (1975) 'Clean Air over London', *Clean Air, 5*, 25-30

Auliciems, A. and Burton, I. (1973) 'Trends in Smoke Concentrations Before and After the Clean Air Act of 1956', *Atmospheric Environment, 7*, 1063-70

Ball, D.J. and Bernard, R.E. (1978) 'An Analysis of Photochemical Pollution Incidents in the Greater London Area with Particular Reference to the Summer of 1976', *Atmospheric Environment, 12*, 1391-401

Ball, D.J. and Hume, R. (1977) 'The Relative Importance of Vehicular and Domestic Emissions of Dark Smoke in Greater London in the mid-1970s', *Atmospheric Environment, 11*, 1065-73

Barnes, R.A. (1979) 'The Long-range Transport of Air Pollution – A Review of European Experience', *J. Air Pollut. Control Ass., 29*, 1219-35

Barnes, R.A. and Lee, D.O. (1978) 'Visibility in London and the Long Distance Transport of Atmospheric Sulphur', *Atmospheric Environment, 12*, 791-4

Bernstein, H.T. (1975) 'The Mysterious Disappearance of Edwardian London Fog', *The London Journal, 1*, 189-206

Berry, B.J.L. (1977) *The Social Burdens of Environmental Pollution: A Comparative Metropolitan Data Source*, Ballinger, Cambridge, Mass.

Berry, B.J.L. and Horton, F.E. (1974) *Urban Environmental Management: Planning for Pollution Control*, Prentice-Hall, New Jersey

Boxer, B. (1981) 'Environmental Science and Policy in China', *Environment, 23*, 14-20, 36-7

Brazell, J.H. (1970) 'Meteorology and the Clean Air Act', *Nature, 226*, 694-6

Brimblecombe, P. (1977) 'London Air Pollution 1500-1900', *Atmospheric Environment, 11*, 1157-62

Brimblecombe, P. (1978) 'London Air Pollution 1500-1900', *Atmospheric Environment, 12*, 2522-3

Brimblecombe, P. (1982) 'Long Term Trends in London Fog', *The Science of the Total Environment, 22*, 19-29

Brodine, V. (1971) 'Episode 104', *Environment, 13*, 2-28

Brodine, V. (1972) 'Running in Place', *Environment, 14*, 2-11, 52

Burton, I., Billingsley, D., Blacksell, M., Chapman, V., Kirkby, A.V., Foster, L. and Wall, G. (1974) 'Public Response to a Successful Air Pollution Control Programme' in Taylor, J.A. (ed.) *Climatic Resources and Economic Activity*, David and Charles, Newton Abbot

Chandler, T.J. (1976) 'Urban Climatology and its Relevance to Urban Design', *WHO Technical Note 149*, World Health Organisation, Geneva, pp. 11-14, 44-7

Chapman, K. (1981) 'Issues in Environmental Impact Assessment', *Prog. in Human Geog., 5*, 191-210

Collin, R.L. (1978) 'The Strategy for Cleaning up our Waters' in Toribara, T.Y., Coleman, J.R., Dahneke, B.E. and Feldman, I. (eds.) *Environmental Pollutants: Detection and Measurement*, Plenum Press, New York, pp. 19-35

Costle, D.M. (1979) 'Dollars and Sense: Putting a Price Tag on Pollution', *Environment, 21*, 25-7

Cox, R.A., Eggleton, A.E.J., Derwent, R.G., Lovelock, J.E. and Pack, D.H. (1975) 'Long-range Transport of Photochemical Ozone in Northwestern Europe', *Nature, 255*, 118-21

Department of Environment (1979) *Digest of Environmental Pollution Statistics*, HMSO, London

Derwent, R.G., McInnes, G., Stewart, H.N.M. and Williams, M.L. (1976) 'The Occurrence and Significance of Air Pollution by Photochemically Produced Oxidant in the British Isles 1972-5', *Report LR 227 (AP)*, Warren Spring Laboratory, Stevenage

Dworkin, J.M. and Pijawka, K.D. (1981) 'Air Quality and Perception: Explaining Change in Toronto, Ontario', *Working Paper EPR-9*, Institute for Environmental Studies, University of Toronto, Toronto

Elsom, D.M. (1978a) 'The Changing Nature of a Meteorological Hazard', *J. of Meteorology, U.K.*, *3*, 297-9

Elsom, D.M. (1978b) 'Meteorological Aspects of Air Pollution Episodes in Urban Areas', *Discussion Papers in Geography 5*, Oxford Polytechnic, Oxford

Elsom, D.M. (1979) 'Air Pollution Episode in Greater Manchester', *Weather*, *34*, 277-86

Elsom, D.M. (1982) 'Drastic Reduction in the United Kingdom Air Pollution Monitoring Network', *Geography*, *67*, 134-7

Greater London Council (1979) *Public Services and Safety Committee Report*, 15 February 1979

Gross, M. (1982) 'Computer Simulation in Urban Planning and Air Pollution Control', *J. Environmental Systems*, *11*, 257-69

Handscombe, C. and Elsom, D.M. (1982) 'Rationalisation of the National Survey of Air Pollution Monitoring Network of the United Kingdom using Spatial Correlation Analysis: A Case-study of the Greater London Area', *Atmospheric Environment*, *16*, 1061-70

Harris, G.R. (1981) 'Positive Impacts of Environmental Policy of Business in the United States', *Intern. J. Environmental Studies*, *16*, 75-83

Harris, B.D. and Smith, K. (1982) 'Cleaner Air Improves Visibility in Glasgow', *Geography*, *67*, 137-9

Hart, G. (1982) 'Clean Air: Time for Responsible Reform', *J. Air Pollut. Control Ass.*, *32*, 14-18

Heneson, N. (1982) 'Environmental Medicine could Kill the Patient', *New Scientist*, *93*, 701

Hewitt, K. and Burton, I. (1971) 'The Hazardousness of a Place: A Regional Ecology of Damaging Events', *Research Publication 6*, Department of Geography, University of Toronto, Toronto

Holister, G. and Porteous, A. (1976) *The Environment: A Dictionary of the World Around Us*, Arrow, London

Huanxuan, G. (1981) 'Environmental Protection in China', *Beijing Review*, *26*, 12-15

Johnson, S.P. (1979) *The Pollution Control Policy of the European Communities*, Graham and Trotman, London

Komarov, B. (1980a) *The Destruction of Nature in the Soviet Union*, Pluto Press, Nottingham

Komarov, B. (1980b) Soviet Conservation: A Bear with no Claws', *New Scientist*, *88*, 514-5

Kormondy, E.J. (1980) 'Environmental Protection in Hungary and Poland', *Environment*, *22*, 31-7

Kromm, D.E. (1973) 'Response to Air Pollution in Ljubljana, Yugoslavia', *AAAG*, *63*, 208-17

Kromm, D.E., Probalb, F. and Wall, G. (1973) 'An International Comparison of Response to Air Pollution', *J. Environmental Management*, *1*, 363-75

Levitt, R. (1980) *Implementing Public Policy*, Croom Helm, London

Lin, G-Y. (1981) 'Simple Markov Chain Model of Smog Probability in the South Coast Air Basin of California', *Prof. Geog.*, *33*, 228-36

McAfee, J. (1982) 'Clean Air, Energy, and Jobs: Can we have them All?', *J. Air*

Pollut. Control Ass.,32, 8–18

McGinty, L. (1977) 'Air Pollution Underestimated in London', *New Scientist*, 75, 141

Marshall, E. (1978) 'EPA Smog Standard Attacked by Industry, Science Advisers', *Science*, *202*, 949–50

Marshall, E. (1979) 'Smog's not so Bad, EPA Decides', *Science*, *203*, 529

Marshall, E. (1981) 'Cleaning up the Clean Air Act', *Science*, *214*, 1328–9

Murchett, F.D. (1981) 'Spatial Distributions of Urban Atmospheric Particulate Concentrations', *AAAG*, *71*, 552–65

National Clean Air Coalition (1981) 'Cleaning up the Clean Air Act: Two Views', *Environment*, *23*, 16, 20, 42–4

National Environmental Development Association (1981) 'Cleaning up the Clean Air Act: Two Views', *Environment*, *23*, 17–20

de Nevers, N.H., Neligan, R.E. and Slater, H.M. (1977) 'Air Quality Management, Pollution Control Strategies, Modelling and Evaluation' in Stern, A.C. (ed.) *Air Pollution*, V, 3rd edn., 3–40

Padmanabhamurty, B. and Hirt, M.S. (1974) 'The Toronto Heat Island and Pollution Distribution', *Water, Air & Soil Pollut.*, *3*, 81–9

Perry, A.H. (1981) *Environmental Hazards in the British Isles*, George Allen and Unwin, London, pp. 128–48

Purdom, P.W. (1980) *Environmental Health*, 2nd edn., Academic Press, New York

Rich, V. (1981) Polish Pollution: 'Smelter Shuts Down', *Nature*, *289*, 112

Rosencranz, A. (1980) 'The Problem of Transboundary Pollution', *Environment*, *22*, 15–20

Rosencranz, A. (1981) 'Economic approaches to Air Pollution Control', *Environment*, *23*, 25–30

Royal Commission on Environmental Pollution (1971) *First Report*, HMSO, London, pp. 11–12

Royal Commission on Environmental Pollution (1976) *Fifth Report: Air Pollution Control – An Integrated Approach*, HMSO, London

Saarinen, T.F. and Cooke, R.U. (1970) 'Public Perception of Environmental Quality in Tucson, Arizona', *Occasional Paper 9*, Department of Geography, University College London, London

Sandbach, F. (1980) *Environment, Ideology and Policy*, Basil Blackwell, Oxford

Scarrow, H.A. (1972) 'The Impact of British Domestic Air Pollution Legislation', *Brit. J. Polit. Science*, *2*, 261–82

Smil, V. (1980a) 'China's Environment', *Current History*, *79*, 14–18

Smil, V. (1980b) 'Environmental degradation in China', *Asia Survey*, *20*, 777–8

Smith, R.J. (1981a) 'EPA and Industry Pursue Regulatory Options', *Science*, *211*, 796–8

Smith, R.J. (1981b) 'The Fight over Clean Air Begins', *Science*, *211*, 1328–30

Stafford, H.A. (1977) 'Environmental Regulations and the Location of US Manufacturing: Speculations', *Geoforum*, *8*, 243–8

Starkie, D.N.M. (1976) 'The Spatial Dimensions of Pollution Policy' in Coppock, J.T. and Sewell, W.R.D. (eds.) *Spatial Dimensions of Public Policy*, Pergamon, Oxford, pp. 148–63

Thornes, J.E. (1977) 'Ozone Comes to London', *Prog. in Physical Geog.*, *1*, 506–17

Thornes, J.E. (1979) 'The Best Practicable Means of Air Quality Management in the European Community?', *Prog. in Physical Geog.*, *3*, 427–42

Timberlake, L. (1981) 'Poland – the Most Polluted Country in the World', *New Scientist*, *92*, 248–50

Tout, D.G. (1979) 'The Improvement in Winter Sunshine Totals in City Centres', *Weather*, *34*, 67–71

Unsworth, M.H., Shakespeare, N.W., Milner, A.E. and Ganendra, T.S. (1979)

'The Frequency of Fog in the Midlands of England', *Weather*, *34*, 72–7

Van Ardsol, M.D., Sabagh, G. and Alexander, F. (1964) 'Reality and the Perception of Environmental Hazards', *J. Health & Human Behavior*, *5*, 144–53

Wall, G. (1973) 'Public Response to Air Pollution in South Yorkshire, England', *Environment & Behavior*, *5*, 219–48

Wall, G. (1974a) 'Complaints Concerning Air Pollution in Sheffield', *Area*, *6*, 3–8

Wall, G. (1974b) 'Public Response to Air Pollution in Sheffield, England', *Intern. J. Environmental Studies*, *5*, 259–70

Wall, G. (1976) 'Air Pollution', *Prog. in Geog.*, *8*, 96–131

Wetstone, G.S. (1980) 'The Need for a New Regulatory Approach', *Environment*, *22*, 9–20

Whittow, J. (1980) *Disasters: The Anatomy of Environmental Hazards*, Allen Lane, London, pp. 367–70

Wood, C.M., Lee, N., Luker, J.A. and Saunders, P.J.W. (1974) *The Geography of Pollution: A Study of Greater Manchester*, Manchester University Press, Manchester

NOTES ON CONTRIBUTORS

Dr P.J. Bull, Department of Geography, Queen's University Belfast, N. Ireland.

Dr R.L. Davies, Department of Geography, University of Newcastle-upon-Tyne, England.

Dr D.M. Elsom, Department of Social Studies, Oxford Polytechnic, England.

Dr J.A. Giggs, Department of Geography, University of Nottingham, England.

Professor D.T. Herbert, Department of Geography, University College Swansea, Wales.

Professor R.J. Johnston, Department of Geography, University of Sheffield, England.

Dr A.M. Kirby, Department of Geography, University of Reading, England.

Dr D.A. Kirby, Department of Geography, St David's University College, Lampeter, Wales.

Dr M. Pacione, Department of Geography, University of Strathclyde, Glasgow, Scotland.

Dr G.C.K. Peach, School of Geography, University of Oxford, England.

Dr S.P. Pinch, Department of Geography, University of Southampton, England.

Mr P.R. White, Transport Studies Group, Polytechnic of Central London, England.

INDEX